Lecture Notes in Mathematics

Edited by A. Dold and B. Eckmann

1261

Herbert Abels

Finite Presentability of S-Arithmetic Groups Compact Presentability of Solvable Groups

Springer-Verlag
Berlin Heidelberg New York London Paris Tokyo

Author

Herbert Abels
Universität Bielefeld, Fakultät für Mathematik
4800 Bielefeld, Federal Republic of Germany

Mathematics Subject Classification (1980): Primary: 20F05, 20G30, 20G25, 22D05, 20F16
Secondary: 20F18, 20G10, 20G35, 11E99, 17B55, 20J05, 22E99

ISBN 3-540-17975-5 Springer-Verlag Berlin Heidelberg New York
ISBN 0-387-17975-5 Springer-Verlag New York Berlin Heidelberg

Printing and binding: Druckhaus Beltz, Hemsbach/Bergstr.
2146/3140-543210

Für Elisabeth

Contents

Introduction

0.1

A group Γ is called *finitely presentable* or *finitely presented* if there is an exact sequence of groups $N \rightarrowtail F \twoheadrightarrow \Gamma$ where F is a finitely generated free group and N is a normal subgroup of F, which is finitely generated as a normal subgroup. There are no general necessary and sufficient criteria for finite presentability. So one has to study certain classes of groups. In this book we study the class of S-arithmetic subgroups Γ of linear algebraic groups G over number fields k and give necessary and sufficient conditions for such groups to have a finite presentation. The conditions are phrased in terms of the structure of G as algebraic group over certain completions of k (Theorem 6.2.4). In case Γ is solvable one can translate these conditions into purely group theoretical conditions (see chapter VII).

0.2 The problem and its solution

0.2.1 I recall the definition of an *arithmetic* and more generally of a S-*arithmetic* group. Let k be a finite extension field of $\mathbb{Q}$ (for the case of positive characteristic see below 0.4.7 ff.). Let G be a *linear algebraic group* defined over k. We think of G as a subgroup of some general linear group GL_n with entries from some big field, where G is given as the set of zeros of a set of polynomials in the variables, $X_{11},\ldots X_{nn}$ and $(\det X)^{-1}$, $X=(X_{ij}) \in GL_n$, with coefficients in k. For any subring R of the big field let $G_R = G \cap GL_{n,R}$, where $GL_{n,R}$

is the group of n×n-matrices with entries from R and determinant a unit of R. For the sake of easier exposition let us take $k = \mathbb{Q}$ in this introduction. A subgroup Γ of G is called *arithmetic* if it is commensurable with $G_{\mathbb{Z}}$, i.e. $\Gamma \cap G_{\mathbb{Z}}$ is of finite index in both Γ and $G_{\mathbb{Z}}$.

For arithmetic groups there is the following definitive result.

0.2.2 THEOREM *(Borel, Harish-Chandra 1962 [19, 15]). Every arithmetic group has a finite presentation.*

A much stronger finiteness result holds for arithmetic groups, see below 0.4.

0.2.3 Next one allows certain denominators in the entries of G. So let $S = \{p_1, \dots p_s\}$ be a finite set of prime numbers and put $\mathbb{Z}(S) = \mathbb{Z}[\frac{1}{p_1}, \dots, \frac{1}{p_s}]$. A *S-arithmetic subgroup* Γ of the linear algebraic group G defined over k is a subgroup of G commensurable with $G_{\mathbb{Z}(S)}$. This book deals with the

0.2.4 PROBLEM *For which G and S is the S-arithmetic subgroup Γ finitely presented?*

The following partial answer has been known.

0.2.5 THEOREM *(Behr 1967 [9]). If G is reductive every S-arithmetic subgroup is finitely presented.*

There is also a stronger finiteness theorem, see below 0.4.

0.2.6 I recall when an algebraic group is called reductive. An endomorphism g of a vector space V is called *unipotent* if g - Id is nilpotent. A subgroup of GL(V) is called unipotent if every one of its elements is unipotent. Every linear algebraic group G has a largest unipotent normal subgroup, called the *unipotent radical* of G and denoted

$rad_u(G)$. Now G is called reductive if $rad_u(G) = \{e\}$. The general linear group GL_n is reductive, e.g.

0.2.7 After Behr's result one is tempted to look at unipotent groups and ask which S-arithmetic groups Γ for unipotent G are finitely presented. The answer is: none, if $S \neq \emptyset$ and $G \neq \{e\}$. In fact every such Γ is not even finitely generated, since Γ has a S-arithmetic subgroup of the following group G_2 as homomorphic image.

$$G_2 = \left\{\begin{pmatrix} 1 & * \\ 0 & 1 \end{pmatrix}\right\} = \{(X_{ij}) \in GL_2 \mid X_{11}=X_{22}=1, X_{21}=0\} .$$

The group $G_{2,\mathbb{Z}(S)}$ is isomorphic to the additive group $\mathbb{Z}(S)$ hence not finitely generated unless $S = \emptyset$.

So for a trigonal group of matrices there must be something non trivial on the diagonal and it must act "strongly" on the unipotent radical if Γ has a chance to be even finitely generated. There is a theorem saying when a S-arithmetic group Γ is finitely generated. We need some preparatory material for stating it. The first step is to pass to the completion.

0.2.8 THEOREM *(Kneser 1964* [36]*) The S-arithmetic subgroup* Γ *of* G *is finitely generated (finitely presented) iff* $G_{\mathbb{Q}_p}$ *is compactly generated (compactly presented) for every* $p \in S$.

The right hand sinde needs explanation. The p-adic completion $\mathbb{Q}_p$ of $\mathbb{Q}$ is a locally compact topological field, so the set $G_{\mathbb{Q}_p}$ of zeros of polynomials with rational coefficients is a closed subgroup of $GL_{n,\mathbb{Q}_p}$, hence a locally compact topological group. So it makes sense to say that it is *compactly generated*, i.e. has a compact set of generators. For the definition of *compact presentability* replace "finite" by "compact" in the definition of finite presentability in 0.1. Here F is the free topological group on some compact set. One can describe compact presentability avoiding the concept of free topological group as follows (see 1.1, [1]).

0.2.9 LEMMA *Let G be a locally compact topological group. G has a compact presentation iff there is an exact sequence* $N \rightarrowtail F \twoheadrightarrow G$ *where F is the (abstract) free group on some compact subset X of G and N is the normal subgroup of F generated by some subset of F with bounded reduced length.*

Kneser's theorem has two advantageous features, firstly one looks at one prime number at a time only and secondly one deals with algebraic groups over local fields $\mathbb{Q}_p$, where more about the structure is known, see e.g. the next theorem.

0.2.10 A linear algebraic group G over a field K is called a K-*split torus* if it is isomorphic over K to a product of copies of GL_1's . G is called K-*split solvable* if it is a semidirect product of a K-split torus T and a unipotent normal subgroup U , $G = T \ltimes U$.

0.2.11 THEOREM *(Borel, Tits [22]) Every linear algebraic group G over a local field K of characteristic zero contains a maximal K-split solvable subgroup H and* G_K/H_K *is compact.*

Since G_K is compactly generated (compactly presented) iff H_K is, our problem 0.2.4 is reduced to the following

0.2.12 PROBLEM *Let G be K-split solvable,* $K = \mathbb{Q}_p$ *. For which G is* G_K *compactly presented?*

The analogous problem for compact generation was solved in [22], see 6.2.5.

0.2.13 THEOREM *Let* $G = T \ltimes U$ *be a K-split solvable group as in 0.2.10* $K = \mathbb{Q}_p$ *. Then* G_K *is compactly generated iff the representation of T on* $\mathfrak{u}^{ab}$ *- the abelianized Lie algebra of U - given by the adjoint representation fixes no non zero vector of* $\mathfrak{u}^{ab}$.

This theorem makes precise the earlier vague statement of 0.2.7 that T act strongly on U .

0.2.14 So for

$$G_3 = \left\{\begin{pmatrix} 1 & * & * \\ 0 & * & * \\ 0 & 0 & 1 \end{pmatrix}\right\} = \{(x_{ij}) \in GL_3 \mid x_{11}=x_{33}=1, x_{ij}=0 \text{ if } i>j\}$$

the S-arithmetic group $\Gamma = G_{3,\mathbb{Z}(S)}$ is finitely generated, since $G_{3,\mathbb{Q}_p}$ is compactly generated (both of which can also be verified directly, without using 0.2.8 and 0.2.13). Is this group finitely presented resp. compactly presented? The answer is no by the following theorem essentially due to Bieri and Strebel.

0.2.15 Let T be a K-split torus and let ρ be a representation of T on the vector space V, both V and ρ defined over K in the category of algebraic groups. The simplest representations are the one-dimensional ones $T \to GL_1$, called *characters*. The characters form a group X(T) with respect to pointwise multiplication, which is free abelian of rank dim T. The representation theory of a split torus is very easy. Every representation space V of T splits into one-dimensional ones. So if we let for every character $\chi \in X(T)$ the *weight space be*

$$V^\chi = \{v \in V \mid \rho(t)v = \chi(t)v \text{ for every } t \in T\},$$

we have

$$V = \oplus V^\chi,$$

where the sum is over all $\chi \in X(T)$. An element $\chi \in X(T)$ is called a *weight* of the representation on V if $V^\chi \neq 0$. We regard X(T) as a lattice in $X(T) \otimes_{\mathbb{Z}} \mathbb{R}$, so all concepts of real linear geometry make sense. A representation is called *tame* if for any two (not necessarily distinct) weights the point 0 is not on the closed segment joining them.

0.2.16 THEOREM *Let* $G = T \ltimes U$ *be a K-split solvable algebraic group as in 0.2.10,* $K = \mathbb{Q}_p$. *If* G_K *has a compact presentation then the representation of* T *on* $\mathfrak{u}^{ab}$ *is tame.*

This theorem is a version for topological groups of a more general theorem of Bieri and Strebel [13, 14] for discrete solvable groups (which also yields directly that $G_{3,\mathbb{Z}(S)}$ is not finitely presented for $S \neq \emptyset$). For G_3 of 0.2.14 the set of weights of T on $\mathfrak{u}^{ab}$ consists of exactly two points $\pm\chi$, $\chi \in X(T)$, as is seen by direct inspection. So $G_{3,\mathbb{Q}_p}$ has no compact presentation.

0.2.17 The necessary condition of tameness in 0.2.16 is not sufficient for compact presentability as is seen by the following example. Let

$$G_4 = \left\{\begin{bmatrix} 1 & * & * & * \\ 0 & * & * & * \\ 0 & 0 & * & * \\ 0 & 0 & 0 & 1 \end{bmatrix}\right\} = \{(X_{ij}) \in GL_4 \mid X_{11}=X_{44}=1, X_{ij}=0 \text{ for } i>j\} .$$

Then $G_{4,\mathbb{Q}_p}$ has a compact presentation. This is an easy consequence of theorem 0.2.18 below (see 5.7.1) or follows from the finite presentability of $\Gamma = G_{4,\mathbb{Z}[\frac{1}{p}]}$ ([3]) and 0.2.8. The group Γ is interesting in group theory because it answers the following question of P. Hall's [31] negatively: Is every homomorphic image of a finitely presented solvable group finitely presented? Equivalently, is every normal subgroup of a finitely presented solvable group finitely generated as a normal subgroup? In Γ even the center is not finitely generated, since it is $Z_{\mathbb{Z}[\frac{1}{p}]} \simeq \mathbb{Z}[\frac{1}{p}]$, where $Z = \{(X_{ij}) \in GL_4 \mid X_{ij}=\delta_{ij} \text{ for } (i,j) \neq (1,4)\}$ is the center of G_4.

The group G_4/Z is an example of a group satisfying tameness - the tameness condition involves only $\mathfrak{u}^{ab}$, so does not notice the difference between G_4 and G_4/Z - but $(G_4/Z)_{\mathbb{Q}_p} = G_{4,\mathbb{Q}_p}/Z_{\mathbb{Q}_p}$ is not compactly presented (1.1.3 b)). So there is no way to decide compact presentability from just looking at the representation of T on $\mathfrak{u}^{ab}$, in general.

The reason that $H_{\mathbb{Q}_p} = (G_4/Z)_{\mathbb{Q}_p}$ has no compact presentation is that the algebraic group H has a non trivial central extension $G \to H$, whose kernel is contained in $(\mathrm{rad}_u\, G)'$. That a group H as in the following theorem has no such extension is equivalent to condition 2) of the following theorem. So I have explained necessity of the two conditions. The theorem says that they are also sufficient (see 6.2).

0.2.18 THEOREM *Let* $H = T \ltimes U$ *be a K-split solvable algebraic group,* $K = \mathbb{Q}_p$ *. Then* H_K *has a compact presentation iff the following two conditions hold.*

1) The representation of T *on* $\mathfrak{u}^{ab}$ *is tame.*

2) Zero is not a weight of the representation of T *on* $H_2(\mathfrak{u})$ *.*

Here $H_2(\mathfrak{u})$ is the second Lie algebra homology of the Lie algebra $\mathfrak{u}$ of U with coefficients in the "big field" regarded as trivial $\mathfrak{u}$-module. It can be computed by the Koszul complex.

Combining 0.2.8, 0.2.11 and 0.2.18 we obtain the following criterion for finite presentability of a S-arithmetic group.

0.2.19 THEOREM *Let* G *be a linear algebraic group defined over* $\mathbb{Q}$ *and let* Γ *be a S-arithmetic subgroup of* G *. Then* Γ *has a finite presentation iff for every* $p \in S$ *the maximal* $\mathbb{Q}_p$*-split solvable subgroup* H *of* G *fulfills conditions 1) and 2) of 0.2.18.*

0.2.20 So in order to apply this criterion one has to find a maximal K-split solvable subgroup H of G . In case G has a maximal torus which is K-split, there is a criterion for compact presentability of G_K in terms of the structure of G itself, avoiding the process of passing to a maximal K-split solvable subgroup (see 6.4).

0.2.21 In case Γ is solvable one can translate the conditions above back into purely group theoretical conditions about Γ as follows (see [8] and chapter VII).

Let Γ be a solvable S-arithmetic group. Then there is a nilpotent normal subgroup Δ of Γ such that Γ/Δ contains a finitely generated abelian subgroup Q of finite index (see 7.5.1).

0.2.22 THEOREM *Let* Γ, Δ, Q *be as above. Then* Γ *has a finite presentation iff the following two conditions hold*

1) Δ^{ab} *is a tame* $\mathbb{Z}Q$*-module.*

2) $H_2(\Delta;\mathbb{Z})$ *is a finitely generated* $\mathbb{Z}\Gamma$*-module.*

0.3 Survey of the proof, contents of the chapters

0.3.1 Note that problem 0.2.12 deals with solvable groups of a special type only. So most of the book deals with solvable topological groups of the following type: Let Q be a finitely generated free abelian group, $\mathfrak{n}$ a Lie algebra over $K = \mathbb{Q}_p$ (or a p-adic field) and let Q act on $\mathfrak{n}$ by Lie algebra automorphisms. We assume that $\mathfrak{n}$ is the sum of weight spaces. We ask

0.3.2 *When has the topological group* $Q \ltimes N$, $N = \exp \mathfrak{n}$, *a compact presentation?*

Note that the group G_K of problem 0.2.12 has a normal subgroup $Q \ltimes N$ as above with compact quotient. So 0.3.2 and 0.2.12 are equivalent problems.

0.3.3 We first study the case that Q is infinite cyclic. The result (2.6.4) is that $Q \ltimes N$ has a compact presentation iff the automorphism α of N corresponding to one of the generators of Q is *contracting* on N , i.e. its powers α^n , $n \in \mathbb{N}$, converge to the map $N \to e$ uniformly on compact subsets of N . This is proved in chapter I (sufficiency) and chapter II (necessity).

0.3.4 For the general case rank $Q \geq 1$ we obtain a necessary condition, called condition 1): The representation of Q on $\mathfrak{n}^{ab}$ is tame (3.2.3). This follows from the necessity part of the case rank $Q = 1$. This necessary condition is not sufficient however, see 0.2.17, 3.3.

0.3.5 In *chapter IV* we try to compose $Q \ltimes N$ of building blocks we can handle already. The building blocks are subgroups of the form $Q \ltimes A$, where A is a subgroup of N on which some element of Q acts contracting. We take the free product of these subgroups amalgamated along their intersections $H = \coprod_{\cap} Q \ltimes A = Q \ltimes \coprod_{\cap} A$. Set $M = \coprod_{\cap} A$. If the representation of Q on $\mathfrak{n}^{ab}$ is tame, it turns out by a complicated computation that M is nilpotent (4.4). Furthermore, if we change H into a topological group H_U locally isomorphic to $Q \ltimes N$, then either the natural map $H_U \to Q \ltimes N$ is an isomorphism, and the very definition of H_U gives a compact presentation of $Q \ltimes N$, or $Q \ltimes N$ has no compact presentation at all. So, supposed the necessary condition 1) holds, compact presentability of $Q \ltimes N$ is equivalent to the non-existence of a certain central extension of N. This may be thought of as a condition about the second homology of the group N.

0.3.6 In *chapter V* this (implicit) condition is transformed into a(n explicit) condition about the second homology of the Lie algebra $\mathfrak{n}$ of N, namely condition 2): "There is no non-zero element in $H_2(\mathfrak{n})$ whose Q-orbit has compact closure". So condition 1) and 2) together are equivalent to compact presentability of $Q \ltimes N$. This is the statement of theorem 5.6.1, the main theorem. To prove it an isomorphism theorem between group homology and Lie algebra homology is needed. In 5.3 such an isomorphism is given explicitly for the second homology. Here the ground field of the Lie algebra is $\mathbb{Q}$. The topology of the groups and Lie algebras involved has to be taken into account to change the ground field to $\mathbb{Q}_p$ (5.4). Finally the ground field is our p-adic field K (5.5).

0.3.7 In *chapter VI* the main theorem is applied to K-split solvable groups to give the criterion for finite presentability of S-arithmetic groups as explained in 0.2 (see 6.2). This involves finding a maximal

K-split solvable subgroup of a linear algebraic group G over K . In 6.3 we give a procedure to determine such a subgroup. In case G contains a maximal torus which is K-split, then one needs not search for a maximal K-split solvable subgroup. We give a criterion for compact presentability of G_K in terms of the structure of G itself (6.4). For the proof of this result we need a theorem (6.6.2) giving a necessary and sufficient condition as to when a k-representation of a Borel subgroup of a k-split reductive group R extends to R . This result may be of independent interest.

0.3.8 The criterion for finite presentability of a S-arithmetic group Γ of chapter VI was phrased in terms of the structure of the ambient linear algebraic group. In *chapter VII* this criterion is translated back into a purely group theoretic condition in case Γ is solvable. One of the results thus obtained is 0.2.22, for other ones see 7.0. In chapter VII the necessary translation machinery is developed.

0.3.9 The contents of chapters III through VII have been briefly reviewed above (0.3.4 - 0.3.8). In *chapter I* the definition of the concepts "compact presentability" and "contracting automorphism" are given and a first link between the two is established (1.3). The results of this section hold for locally compact groups more general than nilpotent Lie groups over p-adic fields and have in fact been used by Behr [10] for questions of compact presentability in positive characteristic.

In *chapter II* we provide results about nilpotent groups and Lie algebras needed later on, especially about generators.

For more information about the contents of the individual chapters see their introduction.

0.4 MORE GENERAL QUESTIONS

0.4.1 The list of finiteness conditions for a group Γ starting with "finitely generated", "finitely presented" can be extended as follows. A group Γ is called of *type* F_n if there is an Eilenberg - Mac Lane complex for Γ with finite n-skeleton. A group Γ is called of *type* FP_n if the trivial $\mathbb{Z}\Gamma$-module $\mathbb{Z}$ has a projective resolution $\ldots \to P_i \to P_{i-1} \to \ldots \to P_0 \to \mathbb{Z}$ with finitely generated $\mathbb{Z}\Gamma$-modules P_i for $i \leq n$. The three properties: FP_1, F_1 and finitely generated are equivalent. The properties F_2 and finitely presented are equivalent and imply FP_2. It is an open problem whether the converse is true. For $n > 2$ a group is of type F_n iff it is finitely presented and of type FP_n ([53, 54]). It is a consequence of this book that a solvable S-arithmetic group is of type FP_2 iff it is finitely presented (see 7.0 Remark 3).

0.4.2 Concerning the question for which G and S the corresponding S-arithmetic group Γ is of type FP_n there are the following results. A group Γ will be called of type FP_∞ resp. F_∞ if it is of type FP_n resp. F_n for every $n \in \mathbb{N}$. We first deal with the number field case $[k:\mathbb{Q}] < \infty$.

0.4.3 THEOREM *Every arithmetic group (i.e.* $S = \emptyset$*) is of type* F_∞.

This was proved for semisimple G by Raghunathan [43] and the general case follows easily (cf. [47] Théorème 4 and [20]).

0.4.4 THEOREM *(Borel, Serre [21]) For reductive* G *every S-arithmetic group is of type* F_∞.

0.4.5 For solvable G the work of Åberg [7] implies that a necessary condition for a S-arithmetic subgroup Γ of G to be of type FP_n is that for every $p \in S$ the representation of the maximal k_p-split subtorus T of $G/\mathrm{rad}_u G$ on the Lie algebra $\mathfrak{u}^{ab}$ of $(\mathrm{rad}_u G)^{ab}$ is n-tame, i.e. zero is not in the convex hull of any $\leq n$ weights of T on $\mathfrak{u}^{ab}$. If $\mathrm{rad}_u G$ is abelian, this condition is also sufficient.

0.4.6 The following sequence of groups

$$G_n = \{(X_{ij}) \in GL_n \mid X_{11} = X_{nn} = 1, X_{ij} = o \text{ for } i > j\}$$

extends our earlier series of examples 0.2.7, 0.2.14, 0.2.17. So one expects that $\Gamma_n = G_{n,\mathbb{Z}[\frac{1}{p}]}$ is of type F_{n-2} but not of type F_{n-1}, which is in fact true ([12, 33, 6, 25]). For $n > 2$ these are the only known examples of S-arithmetic groups of type FP_n with $\mathrm{rad}_u G$ not abelian.

0.4.7 We shall now briefly review the case of a global field of positive characteristic. Let k be a function field over a finite field, $S = \{v_1, \ldots v_s\}$ a finite set of primes of k, $o(S)$ the corresponding *S-arithmetic ring* of all functions holomorphic everywhere except at the points of S, G a linear algebraic group defined over k and Γ a S-arithmetic subgroup of G, i.e. Γ is commensurable with $G_{o(S)}$. Concerning finiteness conditions F_n, there are results known only for reductive G. Let r resp. r_i be the rank (= dimension of a maximal split torus) of G over k resp. k_{v_i}, $v_i \in S$.

0.4.8 Theorem *(Serre [47] Théorème 4) For G reductive and $r = o$, Γ is of type F_∞.*

To state the results for $r > o$ we suppose that G is absolutely almost-simple.

0.4.9 Theorem *Γ is not finitely generated iff $s = r = r_1 = 1$.*

For finite presentability there is the following

0.4.10 CONJECTURE *Γ is not finitely presented iff* $r > 0$ *and*

$$\sum_{i=1}^{s} r_i \leq 2 \;,$$

which has been proved for a number of cases. For more information concerning 0.4.9 and 0.4.10, see Behr [10], in particular what the present state of the art is and to whom the different results are due.

Concerning higher finiteness conditions F_n, $n > 2$, for $r > 0$ there are only the following series of examples.

0.4.11 *(Stuhler [50])* $SL_{2,\mathcal{O}(S)}$ *is of type* F_{s-1} *, not of type* F_s .

0.4.12 *(Abels, Abramenko)* $SL_{n,\mathbb{F}_q[t]}$ *is of type* F_{n-2} *, not of type* F_{n-1} *if* $q \geq \max_{i=1,..n} \binom{n}{i}$.

This result with the bound for q given above is due to Abramenko, Abels proved it only for $q \geq 2^n$ (both unpublished, cf. [5]).

0.5 ACKNOWLEDGEMENTS

In 1983/84 I gave a two semester course on the subject of this book in Bielefeld. In 1985 a preprint was circulated containing chapters I - VI. These chapters are reproduced here with minor changes. The main results of this part are surveyed in [4]. Chapter VII was added in 1986/87.

It is a pleasure to thank all those people who have helped me bringing this book into existence. I thank the listeners of my course in Bielefeld for their patience and helpful questions, in particular A. Gushoff, S. Holz and U. Rehmann. I thank S. Holz for his careful reading of the preprint. I thank H. Behr, R. Bieri, S. Holz, U. Stammbach and R. Strebel for discussions, comments and explanations. I thank my family for their support and encouragement during the preparation of the manuscript. Finally I thank the secretaries C. Draeger, M. Elstner and H. Sternberg for their excellent job of creating this typescript ou of messy manuscripts.

I. COMPACT PRESENTABILITY AND CONTRACTING AUTOMORPHISMS

In this chapter we recall the notion of compact presentability (section 1.1), introduce and discuss contracting automorphisms (section 1.2) and give two results relating the two notions (section 1.3). Our final result relating the two notions will be given only in the next chapter (theorem 2.6.4), after the necessary preparatory material on nilpotent groups.

1.1 COMPACT PRESENTABILITY

In this section we give the definition and two basic properties of compact presentability.

Let A be a group and let X be a subset of A. Denote by $\langle X\rangle$ the subgroup of A generated by X and by $\langle X\rangle^A$ the smallest normal subgroup of A containing X. The set X is called a set of *generators* of A if $\langle X\rangle = A$. Let $R(X\to A)$ be the kernel of the homomorphism $F(X) \to A$ from the free group with basis X to A induced by the inclusion map $X \to A$. The elements of $R(X\to A)$ are called *relations*, or more precisely, relations between elements of X holding in A. A *presentation* $\langle X;R\rangle$ of a group A consists of a subset X of A and a subset R of $R(X\to A)$ such that $\langle X\rangle = A$ and $\langle R\rangle^{F(X)} = R(X\to A)$; the last equation is equivalent to the following two conditions: R is contained in $R(X\to A)$ and the natural homomorphism $F(X)/\langle R\rangle^{F(X)} \to A$ is an isomorphism.

Let A be a locally compact topological group. (Locally compact always is meant to imply Hausdorff.) We call A *compactly generated* if it has a compact set of generators, i.e. there is a compact subset X of A such that $\langle X\rangle = A$. A *compact presentation* $\langle X;R\rangle$ of a locally compact topological group A is a presentation consisting of a compact subset X of A and a compact subset R of the free topological group F(X) with basis X. For the notion of a free topological group see Markov [40]. A locally compact topological group A is called *compactly presentable* or (simpler but less exact) *compactly presented* if A admits a compact presentation. The notion of compact presentability can be reformulated avoiding the notion of a free topological group as follows.

1.1.1 PROPOSITION *Let* A *be a locally compact topological group. The following three conditions are equivalent.*

1) A *has a compact presentation.*

2) There is a compact subset X *of* A *and a number* $n \in \mathbb{N}$ *such that* $\langle X ; \{$*all elements of* $R(X\to A)$ *of reduced length* $\leq n\}\rangle$ *is a presentation of* A.

3) There is a compact subset X *of* A *such that* $\langle X ; \{$*all elements of* $R(X\to A)$ *of reduced length* $\leq 3\}\rangle$ *is a presentation of* A.

The proof of 1)⇒2) follows from the fact that every compact subset of a free topological group has bounded reduced length (s. [28, 29, 45, 2]). For 1)⇒3) s. [1] □

We have the following facts about compact presentability.

1.1.2 PROPOSITION *Let* A *and* B *be locally compact topological groups. Let* $f : A \to B$ *be a continuous proper (i.e. inverse images of compact sets are compact) homomorphism such that* $B/f(A)$ *is compact. Then* A *is compactly generated (compactly presented) iff* B *is.*

The proof of [9, lemma 1] applies, although the lemma stated there is weaker □

1.1.3 PROPOSITION *Let* A *be a closed normal subgroup of a locally compact topological group* B *with quotient* C = B/A.

a) If A *and* C *are compactly generated (compactly presented), so is* B .

b) If B *is compactly generated and* C *is compactly presented, then* A *contains a compact subset* K *such that* $A = \langle K\rangle^B$.

c) If B *is compactly presented and* A *contains a compact subset* K *, such that* $\langle K\rangle^B = A$ *, then* C *is compactly presented.*

For a proof s. [1] □

1.2 CONTRACTING AUTOMORPHISMS

In this section we discuss contracting automorphisms in some detail. Only propostion 1.2.3 will be used later on in this book.

1.2.1 DEFINITION Let N be a locally compact topological group and let α be a topological automorphism of N . The automorphism α is called *contracting (expanding)* if the sequence α^n , $n \in \mathbb{N}$ (α^{-n}, $n \in \mathbb{N}$) of mappings $N \to N$ converges to the map $N \to \{e\}$ uniformly on compact subsets.

1.2.2 REMARK Let N be a locally compact hence finite dimensional, real Banach space with norm $\|\cdot\|$ and let α be an automorphism of the topological group N , hence $\alpha \in GL(N)$. Then α is contracting if there is a real number $C < 1$ such that

$$\|\alpha(x)\| \leq C\|x\|$$

for every $x \in N$. This is the condition in Banach's fixed point theorem. Conversely, if α is contracting in the sense of definition 1.2.1, then there is an equivalent norm on N and a real number $C < 1$ such that this inequality holds, as follows easily from 1.2.3, but there need not be a real number $C < 1$ such that the inequality holds for the given norm.

We are mainly interested in contracting automorphisms of vector spaces and Lie groups over local fields. We shall discuss them now.

For the results on local fields we need see [56, Chapter I]. *A local field* is a commutative locally compact non-discrete topological field. Let K be a local field. There is a natural valuation, the *module* $\mathrm{mod}_K : K \to [0,\infty]$ uniquely determined by the following condition. For every Haar measure μ on the additive group of K and every measurable subset C of K of finite Haar measure $\mu(C)$ we have

$$\mu(x\cdot C) = \mathrm{mod}_K(x) \cdot \mu(C) \quad .$$

The map mod_K is a valuation and defines a topology on K, which is the original topology of K.

Let V be a finite dimensional vector space over K. There is a unique topology on V turning V into a topological vector space over K. In particular, we can regard V as a topological group.

1.2.3 Proposition *Let V be a topological vector space of finite dimension over the local field K. Let α be a K-linear automorphism of V. Let L be a finite extension field of K containing all eigenvalues of α. Then α is contracting on V iff all eigenvalues of α have values less than 1 with respect to mod_L.*

Proof. Let L_2 be an extension field of a local field L_1 of degree d. Then L_2 is a local field. Let w_i be the natural valuation of L_i, $i = 1,2$. Then we have

$$w_2(x) = (w_1(x))^d \quad \text{for every} \quad x \in L_1 \quad ,$$

by Fubini's theorem. So our condition about the values of the eigenvalues of α does not depend on the finite extension field L of K we look at.

Let B be a basis of the vector space V over K . Then α acts contracting on V iff $\alpha^n b$ tends to zero for n to infinity for every $b \in B$. Hence if L is a finite extension field of K , α acts contracting on V iff the L-linear extension of α acts contracting on $V \otimes_K L$. But then the proposition follows from the existence of eigenvectors and the Jordan normal form theorem □

In the next proposition we make use of the notion of a Lie group over a local field K . A Lie group over a loval field is a group and an analytic manifold of finite dimension, such that multiplication and inversion are analytic maps. For results about such Lie groups s. [24 Lie III].

1.2.4 PROPOSITION *Let* N *be a Lie group over a local field* K *of characteristic zero. Let* α *be a contracting automorphism of* N . *Let* β *be the corresponding automorphism of the Lie algebra* $\mathfrak{n}$ *of* N . *Then we have*

a) N *is nilpotent.*

b) β *is a contracting automorphism of* $\mathfrak{n}$.

c) There is a unique exponential map $\Phi : \mathfrak{n} \to N$ *such that* $\alpha \circ \Phi = \Phi \circ \beta$.
This exponential map is an analytic diffeomorphism of $\mathfrak{n}$ *onto* N .

PROOF. Recall that to a Lie group N there is associated a Lie algebra $\mathfrak{n}$ in a functorial way. If K has characteristic zero, there is an exponential maß defined on a neighborhood U of 0 in $\mathfrak{n}$ with image in N . Any two exponential maps coincide on some neighborhood of 0 . Every exponential map is a local analytic diffeomorphism near 0 .

We first prove b). Let $\varphi : U \to N$ be an exponential map for N . We may assume that φ is an analytic diffeomorphism from U to its image $\varphi(U)$. The two exponential maps φ and $\alpha^{-1} \circ \varphi \circ \beta$ coincide on some relatively compact neighborhood $V \subset U$ of 0 in $\mathfrak{n}$, so

$\beta|V = \varphi^{-1}\circ\alpha\circ\varphi|V$. Note that β acts contracting iff some positive power of β acts contracting, as follows easily from the definition. Thus we may assume that $\alpha(\varphi(V)) \subset \varphi(V)$ by passing to some power of α if necessary, since α is contracting. So $\beta(V) \subset V$ and we obtain by induction

$$\beta^n|V = \varphi^{-1}\circ\alpha^n\circ\varphi|V \qquad (*)$$

for $n \in \mathbb{N}$. So $\beta^n|V$ converges uniformly on V to the map $V \to \{e\}$, which implies b), since β is linear.

c) We continue to use the notations and assumptions of the proof of b) . For $X \in \mathfrak{n}$ there is a $n \in \mathbb{N}$ such that $\beta^n(X) \in V$. The element $\alpha^{-n}\circ\varphi\circ\beta^n(X)$ is independent of the number $n \in \mathbb{N}$ by (*). We thus have a map

$$\Phi : \mathfrak{n} \to N$$

well defined by the condition that $\Phi(X) = \alpha^{-n}\circ\varphi\circ\beta^n(X)$, if $\beta^n(X) \in V$. The map Φ has the required properties, the equation $\alpha\circ\Phi = \Phi\circ\beta$ follows easily from the definition and $\Phi|V = \varphi|V$, so Φ is an exponential map. To prove uniqueness let Φ' be an exponential map such that $\alpha\circ\Phi' = \Phi'\circ\beta$. Then there is a neighborhood W of 0 in $\mathfrak{n}$ such that $\Phi|W = \Phi'|W$. For $X \in \mathfrak{n}$ there is a $n \in \mathbb{N}$ such that $\beta^n(X) \in W$, hence $\Phi(X) = \alpha^{-n}\circ\Phi\circ\beta^n(X) = \alpha^{-n}\circ\Phi'\circ\beta^n(X) = \Phi'(X)$. So $\Phi = \Phi'$. The map Φ is an analytic diffeomorphism near 0 . Since $\Phi = \alpha^{-1}\circ\Phi\circ\beta$ and β is contracting, Φ is a local analytic diffeomorphism near every point of $\mathfrak{n}$. The map Φ is surjective, since its image contains a neighborhood of e and is invariant under the expanding map α^{-1} . The map Φ is injective, since if X_1, X_2 are elements of $\mathfrak{n}$, let $n \in \mathbb{N}$ be such that $\beta^n(X_i) \in V$ for $i = 1,2$, then $\alpha^{-n}\Phi\beta^n(X_1) = \Phi(X_1) = \Phi(X_2) = \alpha^{-n}\Phi\beta^n(X_2)$ implies $X_1 = X_2$ since α and β are bijective and $\Phi|V$ is injective.

Now a) follows from the following lemma.

1.2.5 LEMMA *Let* $\mathfrak{n}$ *be a Lie algebra of finite dimension over a local field* K . *If* $\mathfrak{n}$ *admits a contracting Lie algebra automorphism, then* $\mathfrak{n}$ *is nilpotent.*

PROOF. Let β be a contracting automorphism of the K-Lie algebra $\mathfrak{n}$. The claim is proved by looking at the filtration given by the speed of convergence of β , to be defined in a moment.

By extending the ground field, we may assume that K contains all eigenvalues of β . Let $\|\cdot\|$ be a norm on the K-vector space $\mathfrak{n}$. For every positive real number r let $\mathfrak{n}_r$ be the K-subspace of elements X of $\mathfrak{n}$ such that $\left\{\frac{\|\beta^n(X)\|}{r^n};\ n \in \mathbb{N}\right\}$ is bounded. Since the Lie bracket is a bounded bilinear maß $\mathfrak{n} \times \mathfrak{n} \to \mathfrak{n}$, we have

$$[\mathfrak{n}_r , \mathfrak{n}_{r'}] \subset \mathfrak{n}_{r \cdot r'} \qquad (**) \quad .$$

If $r_1 < 1$ is the maximum of the values $\mathrm{mod}_K(\lambda)$ of the eigenvalues λ of β , we have $\mathfrak{n} = \mathfrak{n}_{r_1}$. On the other hand, if λ is an eigenvalue of the restriction of β to $\mathfrak{n}_r$, then $\mathrm{mod}_K(\lambda) \leq r$. So there is a number $r_o > 0$, such that $\mathfrak{n}_{r_o} = 0$. Hence if $r_1^s \leq r_o$, the s-term of the descending central series of $\mathfrak{n}$ is zero by (**) □

1.2.6 COROLLARY *Let* G *be a locally compact topological group whose group* G/G_o *of connected components is compact. Suppose* G *admits a contracting automorphism. Then* G *is a simply connected nilpotent real Lie group.*

PROOF. Note first that a compact topological group admits a contracting automorphism iff it consists of one element. Apply this remark to G/G_o to see that G is connected. Apply it to the maximal compact normal subgroup of the connected group G to see that G has no small normal subgroups, hence is a real Lie group. Now apply the proposition □

The question arises which nilpotent Lie algebras admit a contracting automorphism. Here is the answer for K of characteristic zero.

1.2.7 PROPOSITION *Let $\mathfrak{n}$ be a finite dimensional Lie algebra over the local field K of characteristic 0. There is a contracting K-Lie algebra automorphism of $\mathfrak{n}$ iff $\mathfrak{n}$ admits a gradation, i.e. $\mathfrak{n}$ can be written as a direct sum of vector subspaces $\mathfrak{n}_j$, $j \in \mathbb{N}$, such that $[\mathfrak{n}_j, \mathfrak{n}_k] \subset \mathfrak{n}_{j+k}$.*

PROOF. Suppose $\mathfrak{n} = \bigoplus_{i \in \mathbb{N}} \mathfrak{n}_i$ with $[\mathfrak{n}_j, \mathfrak{n}_k] \subset \mathfrak{n}_{j+k}$. Let λ be a non-zero element of K with $\mathrm{mod}_K(\lambda) < 1$. Define a contracting automorphism β of $\mathfrak{n}$ by $\beta|\mathfrak{n}_j = \lambda^j \cdot \mathrm{Id}_{\mathfrak{n}_j}$ for every $j \in \mathbb{N}$.

Conversely, let β be a contracting automorphism of $\mathfrak{n}$. Let L be a finite Galois extension of K containing all eigenvalues of β. Define for $\lambda \in L$ the β-submodule $\mathfrak{n}_L^{(\lambda)}$ of $\mathfrak{n}_L = \mathfrak{n} \otimes_K L$ by

$$\mathfrak{n}_L^{(\lambda)} = \{X \in \mathfrak{n}_L \; ; \; (\beta - \lambda \mathrm{Id})^m X = 0 \quad \text{for some} \quad m \in \mathbb{N}\} .$$

It is easy to prove by induction on m that $(\beta-\lambda\mathrm{Id})^m X = 0$ and $Y \in \mathfrak{n}_L^{(\mu)}$ implies $[X,Y] \in \mathfrak{n}_L^{(\lambda\cdot\mu)}$, hence

$$[\mathfrak{n}_L^{(\lambda)}, \mathfrak{n}_L^{(\mu)}] \subset \mathfrak{n}_L^{(\lambda\cdot\mu)} .$$

For $r \in \mathbb{R}$ define $\mathfrak{n}_L^{(r)} = \bigoplus_{\mathrm{mod}_L(\lambda)=r} \mathfrak{n}_L^{(\lambda)}$. Then $\mathfrak{n}_L^{(r)}$ is invariant under the Galois group G of L over K. So we obtain $\mathfrak{n} = \bigoplus_r (\mathfrak{n}_L^{(r)} \cap \mathfrak{n})$, for the set of G-fixed points, since for K of characteristic zero we have the projection $|G|^{-1} \sum_{g \in G} g \in K[G]$ of every K[G]-module onto its subspace of G-fixed vectors. Define

$$\mathfrak{n}^{(r)} = \mathfrak{n}_L^{(r)} \cap \mathfrak{n} .$$

We have $\mathfrak{n} = \bigoplus_r \mathfrak{n}^{(r)}$ and $\mathfrak{n}^{(r\cdot r')} \subset \mathfrak{n}^{(r)} \cdot \mathfrak{n}^{(r')}$.

Let R be the set of elements $r \in \mathbb{R}$ such that $\mathfrak{n}^{(r)} \neq 0$. Suppose there is a map f from R to the positive integers such that $f(r_1 \cdot r_2) = f(r_1) + f(r_2)$ if r_1, r_2 and $r_1 \cdot r_2$ are in R. Then

$\mathfrak{n}_j := \bigoplus_{f(r)=j} \mathfrak{n}^{(r)}$ defines a gradation of $\mathfrak{n}$. In order to prove the existence of such a function f, let us look at the $\mathbb{Q}$-vector space $V \subset \mathbb{R}$ spanned by $S = \{-\log r, r \in R\}$. Note that $S \subset (0,\infty)$. So every non-trivial linear combination of S with non-negative integer coefficients is positive. So by corollary A.2 in the appendix there is a $\mathbb{Q}$-linear form $\ell : V \to \mathbb{Q}$ having positive values on S. We may assume that these values are positive integers. Then define $f = \ell \circ -\log$ □

1.2.8 Note that not every nilpotent Lie algebra admits a contracting automorphism. J.L. Dyer [27] has given an example of a nilpotent Lie algebra over $\mathbb{R}$ of dimension 9 every automorphism of which is unipotent.

1.3 COMPACT PRESENTABILITY AND CONTRACTING AUTOMORPHISMS

The notions of compact presentation and contracting automorphism are related by the two results of this paragraph. The first one describes the building blocks for our theory of compact presentability. The second one is a step towards a converse result of the first one. It will be used in the proof of the converse theorem 2.6.4 (or equivalently 3.2.4).

1.3.1 PROPOSITION *Let N be a locally compact topological group. Let α be a contracting automorphism of N. Let $\langle t\rangle \simeq \mathbb{Z}$ act on N by sending t to α. Then the corresponding split extension $\langle t\rangle \ltimes N$ has a compact presentation.*

PROOF. Let K be a compact neighborhood of e in N. We have $\alpha^m(K) \subset K$ for some $m \in \mathbb{N}$. By passing to the open subgroup $\langle t^m\rangle \cdot N$ of finite index we may assume that $m = 1$ (s. 1.1.2). Let $G = \langle t\rangle \cdot N$, let $X = K \cup \{t\}$ and let R be the set of relations in $R(X \to G)$ of reduced length ≤ 4 . I claim that $<X;R>$ is a presentation of G .

Let $H = F(X)/<R>^{F(X)}$. Since $R \subset R(X \to H)$ the map $X \to G$ induces a homomorphism $\pi : H \to G$. I have to show that π is an isomorphism. We show that the natural map $\varphi_o = X \to H$ extends to a homomorphism $\varphi : G \to H$, which is then an inverse of π , since X generates G .

Instead of $\varphi_o(y)$ let us write $\underline{y}$. For $y \in K$ the equation

$$\underline{\alpha(y)} = \underline{t} \cdot \underline{y} \cdot \underline{t}^{-1}$$

holds in H , since $tyt^{-1} = \alpha(y) \in K$, so $\underline{\alpha(y)}^{-1} \cdot \underline{t} \cdot \underline{y} \cdot \underline{t}^{-1}$ is in R . By induction on m we obtain

$$(*) \qquad \underline{\alpha^m(y)} = \underline{t}^m \cdot \underline{y} \cdot \underline{t}^{-m} \quad \text{for} \quad y \in K , \ m \geq 0 \ .$$

For every $u \in N$ there is an integer n such that $\alpha^n(u) \in K$. If for $u \in N$ we have two integers n_1, n_2 with $\alpha^{n_i}(u) \in K$, $i = 1,2$, then

$$\underline{t}^{-n_1} \cdot \underline{\alpha^{n_1}(u)} \cdot \underline{t}^{n_1} = \underline{t}^{-n_2} \cdot \underline{\alpha^{n_2}(u)} \cdot \underline{t}^{n_2} \ .$$

To prove this suppose $n_1 \geq n_2$ and apply $(*)$ with $m = n_1 - n_2$. So there is a well defined map $\varphi_1 : N \to H$ such that

$$\varphi_1(u) := t^{-n} \cdot \alpha^n(u) \cdot t^n \quad \text{if} \quad \alpha^n(u) \in K \ .$$

Obviously φ_1 extends φ_o .

I claim that φ_1 is a homomorphism. Note first that $\underline{y} \cdot \underline{z} = \underline{yz}$ if y,z and yz are in K , because $\underline{y} \cdot \underline{z} \cdot (\underline{yz})^{-1}$ is in R . Now if u_1, u_2 are two elements of N , take an integer n such that $\alpha^n(u_1)$, $\alpha^n(u_2)$ and $\alpha^n(u_1 \cdot u_2)$ are in K . Then apply the preceding remark and the definition of φ_1 to see that $\varphi_1(u_1) \cdot \varphi_1(u_2) = \varphi_1(u_1 \cdot u_2)$.

We have a unique homomorphism $\varphi_2 : <t> \to H$ such that $\varphi_2(t) = \varphi_o(t) = \underline{t}$. Define $\varphi : G = <t> \ltimes N \to H$ by $\varphi(t^n \cdot u) = \varphi_2(t^n) \cdot \varphi_1(u) = \underline{t}^n \cdot \varphi_1(u)$. In order to show that φ is a homomorphism it suffices to show that

$$\underline{t}^m \varphi_1(u) t^{-m} = \varphi_1(t^m u t^{-m})$$

holds for ever $m \in \mathbb{Z}$ and $u \in N$. But this follows from the definition of φ_1 by taking $n \in \mathbb{Z}$ such that $\alpha^n(u)$ and $\alpha^n(t^m u t^{-m}) = \alpha^{n+m}(u)$ are in K □

There is the following weak converse of proposition 1.3.1. It is strong enough, however, to give the full converse of 1.3.1 in the situation we are interested in, see 2.6.4. The analogue of proposition 1.3.2. for discrete groups is due to Bieri and Strebel [13]. I thank them for drawing my attention to this fact. The proof given below is an adaption of theirs.

1.3.2 PROPOSITION *Let* N *be a solvable locally compact topological group, let* α *be a topological automorphism of* N *. Let* $\langle t\rangle \simeq \mathbb{Z}$ *act on* N *by sending* t *to* α *. If the corresponding split extension* $\langle t\rangle \ltimes N$ *has a compact presentation, then there is an open compactly generated subgroup* B *of* N *such that 1)* $\alpha^{-1}(B) \subset B$ *and* $\bigcup_{n\in\mathbb{N}} \alpha^n B = N$ *or 2)* $\alpha(B) \subset B$ *and* $\bigcup_{n\in\mathbb{N}} \alpha^{-n} B = N$.

PROOF. If $G = \langle t\rangle \ltimes N = \langle t\rangle \cdot N$ has a compact presentation, let $\bigcup_{i\in\mathbb{Z}} t^i A_i$ be a compact set of generators of G , where the A_i are compact subsets of N , empty for all but a finite set of integers i . Hence there is a compact subset A of N , such that $A \cup \{t\}$ generates G , e.g. $A = \bigcup A_i$. We may assume that A contains a neighborhood of e and is symmetric, i.e. $A = A^{-1}$. Let V be the interior of A^2 . We have $V = V^{-1} \supset A$.

We shall define a group H by giving a presentation. Let $X = V \cup \{t\}$ and let $\varphi : X \to F(X)$ be the natural map of X into the free group $F(X)$ with basis X . Let R_1 be the set of all relations in $R(X\to G)$ of reduced length ≤ 3 and let R_2 be the set of relations of the form

$$\varphi(t) \cdot \varphi(v) \cdot \varphi(t)^{-1} \cdot \varphi(\alpha(v))^{-1}$$

with $v \in V$ and $\alpha(v) \in V$. Set $R = R_1 \cup R_2 \subset R(X\to G)$. Define H as the group with presentation $\langle X;R\rangle$, i.e. $H = F(X)/\langle R\rangle^{F(X)}$. Let

$X \to H$, $x \longrightarrow \underline{x}$, be the obvious map. The group H has a unique topology such that H is a topological group and the map $V \to H$, $v \longrightarrow \underline{v}$, is a homeomorphism onto an open subset of H . The proof is similar to that in [52 p. 58 ff]. So the natural homomorphism $\pi : H \to G$ is an open continuous surjective homomorphism with discrete kernel D . Now H has a compact set of generators, namely $\{\underline{x} ; x \in A \cup \{t\}\}$, and G has a compact presentation. So by proposition 1.1.3 there is a finite subset P of D such that $\langle P\rangle^H = D$. Now D is contained in the kernel of $H \to G \to \langle t\rangle$, which is generated as a group by $\cup\, \underline{t}^j \underline{A}\, \underline{t}^{-j}$, $j \in \mathbb{Z}$. So there are integers m and n with $m < 0 < n$ such that $P \subset \left\langle \cup\, \underline{t}^j \cdot \underline{A} \cdot \underline{t}^{-j} , m \le j \le n \right\rangle$. Define

$$B = \left\langle \cup\, t^j \cdot A \cdot t^j , m \le j \le n \right\rangle$$
$$S = \left\langle \cup\, t^j \cdot A \cdot t^{-j}, m \le j \le n-1 \right\rangle$$
$$T = \left\langle \cup\, t^j \cdot A \cdot t^{-j}, m+1 \le j \le n \right\rangle$$

and let $\beta : S \xrightarrow{\sim} T$ be the isomorphism induced by conjugation with t . Let G^* be the following group. Take the free product of B with an infinite cyclic group $\langle y\rangle$ and mod out the relations $y \cdot x \cdot y^{-1} = \beta(x)$ for $x \in S$. So G^* is the HNN-extension $G^* = B*_\beta$. We have a unique homomorphism $\Psi : H \longrightarrow G^*$ with $\Psi(a) = a$ for $a \in A$ and $\Psi(\underline{t}) = y$ and an obvious homomorphism $G^* \longrightarrow G$ sending y to t . Both are surjective and their composition is $\pi : H \longrightarrow G$. Since by construction $\langle P\rangle^H = D = \ker \pi$ is contained in $\ker \Psi$, the map $G^* \to G$ is an isomorphism. So G is an HNN-extension $G = B*_\beta$ with $\beta : S \xrightarrow{\sim} T$. Now it is well known that such an HNN-extension contains a non abelian free subgroup, hence is not solvable, unless $B = S$ or $B = T$ (see [35]). In the first case $tBt^{-1} = tSt^{-1} \subset B$, so $\bigcup_{j \ge 0} t^j B t^{-j} = B$ and the ascending chain $t^{-j} B t^j$, $j \ge 0$, of subgroups of N has union N , since N is generated by $\cup\, t^j A t^{-j}$, $j \in \mathbb{Z}$. Similarly $N = \bigcup_{j \ge 0} t^j B t^{-j}$ in the second case.

II. Filtrations of Lie algebras and groups

After the necessary preparatory material on nilpotent Lie algebras and Lie groups over fields of characteristic zero we discuss generating sets for such Lie groups in 2.5. We also prove that a compactly generated subgroup of a nilpotent p-adic Lie group has compact closure (2.6.3). This implies a converse 2.6.4 of 1.3.1, relating compact presentability and contracting automorphisms.

2.1 Filtered Lie algebras

In this section the notions of filtration, descending central series and nilpotence of Lie algebras are recalled. Note proposition 2.1.3.

Let $\mathfrak{g}$ be a Lie algebra over some sommutative ring R. If $\mathfrak{a}$ and $\mathfrak{b}$ are ideals of $\mathfrak{g}$, we denote by $[\mathfrak{a},\mathfrak{b}]$ the $\mathbb{Z}$-submodule of $\mathfrak{g}$ generated by the set of Lie brackets $[X,Y]$, $X \in \mathfrak{a}$, $Y \in \mathfrak{b}$. It is an R-submodule of $\mathfrak{g}$ and the Jacobi identity implies that $[\mathfrak{a},\mathfrak{b}]$ is an ideal of $\mathfrak{g}$.

A *filtration* of $\mathfrak{g}$ is a decreasing sequence $\mathfrak{g}_i$, $i \in \mathbb{N}$, of R-submodules of $\mathfrak{g}$ such that

$$\mathfrak{g} = \mathfrak{g}_1$$

and

(2.1.1) $$[X,Y] \in \mathfrak{g}_{i+j} \quad \text{for} \quad X \in \mathfrak{g}_i \quad \text{and} \quad Y \in \mathfrak{g}_j \ .$$

It follows that $[\mathfrak{g}_i,\mathfrak{g}_j] \subset \mathfrak{g}_{i+j}$, the $\mathfrak{g}_i$'s are ideals of $\mathfrak{g}$ and $\mathfrak{g}_{i-1}/\mathfrak{g}_i$ is central in $\mathfrak{g}/\mathfrak{g}_i$.

A Lie algebra together with a filtration is called a *filtered Lie algebra*. Define the *associated graded Lie algebra* of the filtered Lie algebra $\mathfrak{g} = \mathfrak{g}_1 \supset \mathfrak{g}_2 \supset \ldots$ by

$$\mathrm{gr}\ \mathfrak{g} = \bigoplus_{i=1} \mathrm{gr}_i\ \mathfrak{g}$$

with

$$\mathrm{gr}_i \mathfrak{g} = \mathfrak{g}_i/\mathfrak{g}_{i+1} \quad .$$

Its Lie bracket $\mathrm{gr}_i\ \mathfrak{g} \times \mathrm{gr}_j\ \mathfrak{g} \longrightarrow \mathrm{gr}_{i+j}\ \mathfrak{g}$ is induced by the Lie bracket in $\mathfrak{g}$. Note that $\mathrm{gr}\ \mathfrak{g}$ is a graded Lie algebra over R .

For every Lie algebra we have the filtration by the *descending central series* defined by induction on i as follows. $\mathfrak{g}_1 := \mathfrak{g}$, $\mathfrak{g}_{i+1} := [\mathfrak{g}_i,\mathfrak{g}]$. The Jacobi identity implies that $[\mathfrak{g}_i,\mathfrak{g}_j] \subset \mathfrak{g}_{i+j}$, i.e. that the descending central series is a filtration. Obviously, for every filtration $\mathfrak{g} = \mathfrak{g}_{(1)} \supset \mathfrak{g}_{(2)} \supset \ldots$ we have $\mathfrak{g}_{(i)} \supset \mathfrak{g}_i$. So the descending central series is the fastest decreasing filtration.

A Lie algebra $\mathfrak{g}$ is called *m-step nilpotent* if in the descending central series the m+1-st term $\mathfrak{g}_{m+1}$ is zero, but $\mathfrak{g}_m \neq 0$. The Lie algebra $\mathfrak{g}$ is called *nilpotent*, if it is m-step nilpotent for some $m \in \mathbb{N}$.

Also recal the following notation. For a Lie algebra $\mathfrak{g}$ the *derived Lie algebra* $\mathfrak{g}'$ is by definition $\mathfrak{g}' = [\mathfrak{g},\mathfrak{g}]$. The corresponding *abelianized Lie algebra* of $\mathfrak{g}$ is $\mathfrak{g}^{ab} = \mathfrak{g}/\mathfrak{g}'$. So $\mathfrak{g}^{ab} = \mathrm{gr}_1\mathfrak{g}$ for the descending central series.

2.1.2 REMARK *Let* $\mathfrak{g}$ *be a Lie algebra over* R . *The associated graded Lie algebra of the descending central series is generated - qua Lie algebra over* $\mathbb{Z}$ *- by* $\mathrm{gr}_1\mathfrak{g} = \mathfrak{g}^{ab}$.

PROOF. By definition of the descending central series, $\mathrm{gr}_{i+1}\mathfrak{g}$ is generated as $\mathbb{Z}$-module by the image of the Lie bracket $gr_1\mathfrak{g} \times gr_i\mathfrak{g} \to gr_{i+1}\mathfrak{g}$. By induction on i we obtain that every Lie subalgebra over $\mathbb{Z}$ of $gr\,\mathfrak{g}$ containing $gr_1\,\mathfrak{g}$ contains $gr_i\,\mathfrak{g}$ for every $i \in \mathbb{N}$ □

The following result is one of a whole series of results of the following type. The relevant information about a nilpotent Lie algebra $\mathfrak{g}$ is contained in its "abelian head" $\mathfrak{g}^{ab}$.

2.1.3 PROPOSITION *Let* $\mathfrak{h}$ *and* $\mathfrak{g}$ *be Lie algebras over* R. *Suppose* $\mathfrak{g}$ *is nilpotent. A* R*-Lie algebra homomorphism* $\mathfrak{h} \to \mathfrak{g}$ *is surjective iff the induced map* $\mathfrak{h}^{ab} \to \mathfrak{g}^{ab}$ *is surjective.*

PROOF. Necessity is clear. To prove sufficiency, suppose $\mathfrak{g}$ is m-step nilpotent. Let $\mathfrak{h}_i$ and $\mathfrak{g}_i$ be the i-th term of the descending central series of $\mathfrak{h}$ and $\mathfrak{g}$ resp. Let $\bar{f} : gr\,\mathfrak{h} \to gr\,\mathfrak{g}$ be the map of the associated graded Lie algebras induced by f. If $gr_1\mathfrak{h} = \mathfrak{h}^{ab} \to \mathfrak{g}^{ab} = gr_1\mathfrak{g}$ is surjective, so is $\bar{f}$ by remark 2.1.2. Thus we obtain an exact sequence

$$0 \to f(\mathfrak{h}_i) \cap \mathfrak{g}_{i+1} \to f(\mathfrak{h}_i) \to gr_i\mathfrak{g} \to 0$$

for every $i \leq m$. Comparison with the exact sequences

$$0 \to \mathfrak{g}_{i+1} \to \mathfrak{g}_i \to gr_i\mathfrak{g} \to 0$$

implies by induction on $j = 0,1, \ldots, m-1$ that $f(\mathfrak{h}_{m-j}) = \mathfrak{g}_{m-j}$, which yields our claim for $j = m-1$ □

2.1.4 Given a filtered Lie algebra $\mathfrak{g} = \mathfrak{g}_1 \supset \mathfrak{g}_2 \supset \ldots$ there is a unique topology on $\mathfrak{g}$ such that 1) $\mathfrak{g}$ is a topological group with respect to addition and 2) the set $\{\mathfrak{g}_n, n = 1,2,\ldots\}$ is a neighborhood base of 0. This topology is called the *topology given by the filtration*. All notions and constructions for topological groups now make sense, e.g. Hausdorff, complete, completion.

2.2. FORMULAE ON COMMUTATORS

Before recalling the notions of filtered group, associated graded Lie algebra, nilpotence etc. for groups we have to recall the following formulae on commutators, cf. [46]. Especially noteworthy is identity 2.2.3,3), an analogue of the Jacobi identity, attributed to P. Hall [30], see [38, 57].

Let G be a group, let x,y be elements of G. We will use the following notations

(2.2.1) $$x^y = y^{-1}xy$$

(2.2.2) $$(x,y) = x^{-1}y^{-1}xy \quad .$$

(x,y) is called the *commutator* of x and y. The map $x \to x^y$ is an automorphism of G and we have $x^{yz} = (x^y)^z$.

2.2.3 PROPOSITION *We have the identities*

1) $xy = yx^y = yx(x,y)$, $x^y = x(x,y)$, $(x,x) = 1$, $(y,x) = (x,y)^{-1}$

2) $(x,yz) = (x,z) \cdot (x,y)^z$

2') $(xy,z) = (x,z)^y \cdot (y,z)$

3) $(x^y,(y,z)) \cdot (y^z,(z,x)) \cdot (z^x,(x,y)) = 1$

PROOF. 1) is trivial.

2) From 2.2.1 and 1) we have

$x(x,yz) = y^{yz} = (x^y)^z = [x(x,y)]^z = x^z(x,y)^z = x(x,z)(x,y)^z$,

therefore $(x,yz) = (x,z)(x,y)^z$.

2') $xy(xy,z) = (xy)^z = x^zy^z = x(x,z)y(y,z) = xy(x,z)^y(yz)$,

therefore $(xy,z) = (x,z)^y(y,z)$.

3) $(x^y,(y,z)) = y^{-1}x^{-1}yz^{-1}y^{-1}zyy^{-1}xyy^{-1}z^{-1}yz$

$= y^{-1}x^{-1}yz^{-1}y^{-1}zxz^{-1}yz$.

Put $u = zxz^{-1}yz$,

$v = xyx^{-1}zx$

$w = yzy^{-1}xy$.

Then $$(x^y,(y,z)) = w^{-1}u$$

Analogously, by cyclic permutation

$$(y^z,(z,x)) = u^{-1}v$$
$$(z^x,(x,y)) = v^{-1}w$$

Hence $(x^y,(y,z))(y^z,(z,x))(z^x,(x,y)) = 1$ □

2.3 Filtered groups

In this section we recall the notions of filtration, associated graded Lie algebra, descending central series and nilpotence, this time for groups.

2.3.1 If A and B are subgroups of a group G, denote by (A,B) the subgroup of G generated by the set of commutators (a,b), $a \in A$, $b \in B$. If A,B and C are normal subgroups of G, then so is (A,B) and we have

$$(A,(B,C)) \subset (B,(C,A)) \cdot (C,(A,B)) .$$

The first claim is obvious and the second one follows from 2.2.3,3).

2.3.2 A *filtration* of a group G is a decreasing sequence of subgroups G_i, $i \in \mathbb{N}$, such that

$$(G_i,G_j) \subset G_{i+j} \quad \text{for } i,j \text{ in } \mathbb{N} .$$

So the subgroups G_i are normal in G and G_{i-1}/G_i is central in G/G_i. A group G together with a filtration of G will be called a *filtered group*.

2.3.3 Given a filtered group $G = G_1 \supset G_2 \supset \ldots$ one defines the *associated graded Lie algebra* as follows. Define the $\mathbb{Z}$-modules

$$gr_iG = G_i/G_{i+1} ,$$

$$gr\, G = \bigoplus_{i=1}^{\infty} gr_iG \quad .$$

For $x \in G_i$, $y \in G_j$ the element $(x,y)G_{i+j+1} \in gr_{i+j}G$ depends only on the classes $xG_{i+1} \in gr_iG$ and $yG_{j+1} \in gr_jG$, by 2.2.3,2) and 2'). We thus define a Lie bracket

$$gr_iG \times gr_jG \longrightarrow gr_{i+j}G$$

by

$$[xG_{i+1}, yG_{j+1}] := (x,y)G_{i+j+1} \quad .$$

The Lie bracket is $\mathbb{Z}$-bilinear by 2.2.3,2) and 2'). The Jacobi identity is satisfied by 2.2.3,3). So $gr\, G$ is a Lie algebra over $\mathbb{Z}$. It is called the associated graded Lie algebra of the filtered group G .

2.3.4 Let G be a group. The *descending central series* G_i , $i \in \mathbb{N}$, of G is defined inductively by $G_1 = G$, $G_{i+1} = (G,G_i)$. If follows from 2.3.1 that every G_i is normal in G , by induction on i , and that $(G_i,G_j) \subset G_{i+j}$, by induction on j . In other words, the descending central series is a filtration. For every filtration $G = G_{(1)} \supset G_{(2)} \supset \ldots$ we have $G_i \subset G_{(i)}$ for every $i \in \mathbb{N}$, so the descending central series is the fastest decreasing filtration.

2.3.5 Let G be a group. As usual we denote by G' the *derived group* $G' = (G,G)$ and by G^{ab} the corresponding *abelianized group*

$$G^{ab} = G/G' = G/(G,G) \quad .$$

2.3.6 *For the descending central series of* G *the associated graded Lie algebra is generated - as a Lie algebra over* $\mathbb{Z}$ *- by* $gr_1G = G^{ab}$. This follows immediately form the definitions, cf. 2.1.2.

2.3.7 A group is called *m-step nilpotent*, if for the descending central series we have $G_{m+1} = 1$ but $G_m \neq 1$. G is called *nilpotent*, if it is m-step nilpotent for some $m \in \mathbb{N}$.

2.3.8 PROPOSITION *Let H and G be groups. Suppose G is nilpotent. A homomorphism $f : H \to G$ is surjective iff the induced map $H^{ab} \to G^{ab}$ of abelianized groups is surjective.*

Same proof as for Lie algebras, see 2.1.3 □

2.4 SOME RESULTS ABOUT NILPOTENT GROUPS

We recall here some results about nilpotent groups with proof (see [55] except for 2.4.4). The proofs of all of them make use of the associated graded Lie algebra of the descending central series. We shall apply these results at various places.

2.4.1 DEFINITION Let P be a set of prime numbers. A *P-number* is an integer whose prime factor decomposition contains only primes $\in P$. A group G is called *P-radicable* if for every P-number m and every $x \in G$ there is an element $y \in G$ such that $y^m = x$. The group is called *radicable* if it is P-radicable for the set P of all prime numbers. For G abelian P-radicability is usually called *P-divisibility*.

2.4.2 PROPOSITION *A nilpotent group G is P-radicable iff G^{ab} is P-radicable.*

PROOF. If G is P-radicable, so is G^{ab} , since every homomorphic image of a P-radicable group is P-radicable. Conversely, suppose G^{ab} is P-radicable. Let gr G be the associated graded Lie algebra of the descending central series $G = G_1 \supset G_2 \supset \dots$ of G . By induction on k one proves that the group $gr_k G$ is P-radicable since the Lie bracket $gr_1 G \otimes gr_{k-1} G \to gr_k G$ is surjective and $gr_1 G$ is P-radicable. It follows by induction that G/G_k is P-radicable for every $k \in \mathbb{N}$, since $G_{k-1}/G_k \simeq gr_{k-1}G$ is central in G/G_k. If G is nilpotent, this implies that G is P-radicable □

2.4.3 PROPOSITION AND DEFINITION *Let P be a set of prime numbers. Let G be a nilpotent group and let A be a subgroup of G. The set* $\{x \in G \,;\, x^m \in A$ *for some* P-*number* $m\}$ *is called the* P-*isolator of* A.

a) The P-*isolator of* A *is a subgroup of* G.

b) G *is the* P-*isolator of* A *iff* G^{ab} *is the* P-*isolator of the image of* A *in* G^{ab}.

PROOF. We first show b). Let $G = G_1 \supset G_2 \subset \ldots$ be the descending central series of G and let $gr\,G$ be the associated graded Lie algebra. Define a filtration of A by $A_k = A \cap G_k$. The associated graded Lie algebra $gr\,A$ of this filtration is embedded in $gr\,G$. If G is the P-isolator of A, then gr_1G is the P-isolator of gr_1A. Suppose conversely, that gr_1G is the P-isolator of gr_1A. Then, by induction, gr_kG is the P-isolator of gr_kA, since gr_kG is the image of the $\mathbb{Z}$-bilinear Lie bracket $gr_1G \times gr_{k-1}G \longrightarrow gr_kG$ and the image of $gr_1A \times gr_{k-1}A$ is contained in gr_kA. It follows by induction that G/G_k is the P-isolator of A/A_k, since G_{k-1}/G_k is central in G/G_k. This proves b) since G is nilpotent.

To prove a) we may assume that G is generated by the P-isolator of A. Then G^{ab} is generated by the P-isolator of the image B of A in G^{ab}. Therefore G^{ab} is the P-isolator of B, since G^{ab} is abelian. Then G is the P-isolator of A by b), which proves our claim □

2.4.4 PROPOSITION *Let* G *be a nilpotent group and let* A *be a subgroup of* G. *Let* B *be a subgroup of* G *generated by* A *and a finite subset of the* P-*isolator of* A. *Then* A *is of finite index in* B *and the index* $\#(B/A)$ *is a* P-*number.*

PROOF. We may assume that $G = B$. Then G is the P-isolator of A , by 2.4.3 b). We have to show that A is of finite index in G and $\#G/A$ is a P-number. Again, let $G = G_1 \supset G_2 \supset \dots$ be the descending central series of G , define $A_i = A \cap G_i$ and let $\operatorname{gr} G$ and $\operatorname{gr} A$ be the resp. associated graded Lie algebras. Then $\operatorname{gr}_1 G = G^{ab}$ is generated by $\operatorname{gr}_1 A$ and a finite number of elements of the P-isolator of $\operatorname{gr}_1 A$, so $\operatorname{gr}_1 G/\operatorname{gr}_1 A$ is a finite group of order a P-number. By induction on k it follows that $\operatorname{gr}_k G/\operatorname{gr}_k A$ is a finite group of order a P-number, since it is the image of the map induced by the Lie bracket $\operatorname{gr}_{k-1} G/\operatorname{gr}_{k-1} A \otimes \operatorname{gr}_1 G/\operatorname{gr}_1 A \longrightarrow \operatorname{gr}_k G/\operatorname{gr}_k A$. It follows now by induction that $\#G_k/A_k$ is a finite P-number □

2.4.5 PROPOSITION *Let* G *be a P-radicable nilpotent group. The group of P-torsion elements, i.e. the P-isolator of the subgroup* $\{e\}$ *, is contained in the center of* G .

PROOF. Note that the set of P-torsion elements forms a group by 2.4.3 a). In this proof we shall also have to consider the ascending central series $Z_i(G)$ of G defined as follows. Let $Z(A)$ denote the center of a group A . Then define $Z_o(G) = \{e\}$. If $Z_i(G)$ is a normal subgroup of G , define $Z_{i+1}(G)$ as the inverse image of the center $Z(G/Z_i(G))$ of $G/Z_i(G)$ under the natural homomorphism $G \longrightarrow G/Z_i(G)$. All $Z_i(G)$ are normal subgroups of G . The commutator map $G \times Z_i(G) \to Z_{i-1}(G)$ induces a bilinear map $G^{ab} \times Z_i(G)/Z_{i-1}(G) \to Z_{i-1}(G)/Z_{i-2}(G)$, by 2.2.3, which is non-degenerate in the second variable for $i \geq 2$, i.e. given $z \neq 0$ in $Z_i(G)/Z_{i-1}(G)$ there is an element $\bar{g} \in G^{ab}$ such that $(\bar{g},z) \neq 0$ in $Z_{i-1}(G)/Z_{i-2}(G)$, by definition of $Z_{i-1}(G)$.

If now G is P-radicable, so is G^{ab} . If A is the group of P-torsion elements of G , then the image A_i of $A \cap Z_i(G)$ in $Z_i(G)/Z_{i-1}(G)$ is a P-torsion group. So the bilinear map above restricts to the zero map on $G^{ab} \times A_i \to Z_{i-1}(G)/Z_{i-2}(G)$, since G^{ab}

is P-radicable and A_i is P-torsion. Therefore $A_i = 0$ for $i \geq 2$ by non-degeneracy, i.e. $A \subset Z_1(G) = Z(G)$ □

2.5 NILPOTENT LIE GROUPS OVER FIELDS OF CHARACTERISTIC ZERO

In this section we define a nilpotent group, called a Lie group, corresponding to every nilpotent Lie algebra over a field k of characteristic zero by means of the Campbell-Hausdorff-formula. We show that the exponential map carries the descending central series of the nilpotent Lie algebra to the descending central series of the corresponding Lie group. This implies various results, which compare generators of Lie algebras and of corresponding groups. We shall use these comparison results at several places in the following chapters. Note that the Lie groups in this section have no manifold structure not even a topology.

I recall the properties of the Campbell-Hausdorff formula (cf. [46]). Let X be a set, let L_X be the free Lie algebra on X over $\mathbb{Q}$. A gradation on L_X is defined as follows. L_X^1 is spanned by X and L_X^{n+1} is spanned by the Lie brackets $[y,x]$, $y \in L_X^n$, $x \in X$. Set

$$\hat{L}_X = \prod_{n=1}^{\infty} L_X^n .$$

$\hat{L}_X$ is the completion of L_X with respect to the filtration of L_X given by $L_{X,n} = \sum_{m \geq n} L_X^m$ (s. 2.1.4). The *Campbell-Hausdorff formula* is a certain element

$$x \circ y \in \hat{L}_{\{x,y\}} .$$

Its first three homogeneous components are

$$\text{(2.5.1)} \quad \begin{cases} (x \circ y)^1 = x + y \\ (x \circ y)^2 = \frac{1}{2}[x,y] \\ (x \circ y)^3 = \frac{1}{12}[x,[x,y]] + \frac{1}{12}[y,[y,x]] . \end{cases}$$

In order to state properties of the Campbell-Hausdorff formula we need to be able to apply it to arbitrary elements of $\hat{L}_X$. More generally, let L be a complete Hausdorff (see 2.1.4) filtered Lie algebra over $\mathbb{Q}$, let a and b be elements of L . The map $x \longrightarrow a$, $y \longrightarrow b$ extends to a unique continuous Lie algebra homomorphism $\varphi : \hat{L}_{\{x,y\}} \to L$. Define the Campbell-Hausdorff product of a and b by

$$a \circ b = \varphi(x \circ y) \quad .$$

The following identities hold for a,b,c in L

(2.5.2) $$a \circ 0 = a = 0 \circ a \; ,$$

(2.5.3) $$a \circ (-a) = 0 = -a \circ a \; ,$$

(2.5.4) $$a \circ (b \circ c) = (a \circ b) \circ c \quad .$$

Now let k be a field of characteristic zero, let $\mathfrak{n}$ be a nilpotent Lie algebra over k . We may regard $\mathfrak{n}$ as a complete Hausdorff filtered Lie algebra over $\mathbb{Q}$ by restricting scalars to $\mathbb{Q}$ and considering the descending central series of $\mathfrak{n}$ (as Lie algebra over $\mathbb{Q}$ or k , there is no difference). So we may apply the above to $\mathfrak{n}$. The Campbell-Hausdorff formula defines a polynomial map

$$\mathfrak{n} \times \mathfrak{n} \to \mathfrak{n} \; , \; (X,Y) \longmapsto X \circ Y \; ,$$

which turns $\mathfrak{n}$ into a group N . Its identity element is 0 , by 2.5.2, and the inverse of X is $-X$, by 2.5.3. We call N the *Lie group* corresponding to $\mathfrak{n}$. Note that this notion differs from the usual one, because here N has no analytic or even topological structure. It is just an abstract group defined by the Campbell-Hausdorff multiplication. For the purpose of distinction we sometimes call $N = \exp \mathfrak{n}$ a *Lie group over* k , if $\mathfrak{n}$ is a nilpotent Lie algebra over k . Note also that $\mathfrak{n}$ may be infinite dimensional.

We shall denote the identity map of $\mathfrak{n}$ by $\exp : \mathfrak{n} \to N$ when regarded as a map from the Lie algebra $\mathfrak{n}$ to the Lie group N . Its inverse will be denoted $\log : N \to \mathfrak{n}$.

The Campbell-Hausdorff formula has the form

(2.5.5) $$X \circ Y = X + Y + \tfrac{1}{2}[X,Y] + \dots ,$$

hence the group commutator has the form

(2.5.6) $$(X,Y) = [X,Y] + \dots .$$

The dots in 2.5.5 and 2.5.6 denote linear combinations of Lie monomials in X and Y of degree ≥ 3 with rational coefficients, i.e. $(X,Y) - [X,Y] \in \varphi(\hat{L}_{\{x,y\},3})$ in the notations above, and similarly with 2.5.5.

A Lie algebra homomorphism $f : \mathfrak{n} \to \mathfrak{m}$ may be regarded as a homomorphism of the corresponding Lie groups $f : \exp \mathfrak{n} \to \exp \mathfrak{m}$. In particular, let $\mathfrak{a}$ be an ideal of $\mathfrak{n}$ and let $f : \mathfrak{n} \to \mathfrak{n}/\mathfrak{a}$ be the natural homomorphism. Then $\exp \mathfrak{a}$ is a normal subgroup of $\exp \mathfrak{n}$, because it is the kernel of $f : \exp \mathfrak{n} \to \exp \mathfrak{n}/\mathfrak{a}$. For every $X \in \mathfrak{n}$ we have $\exp X \cdot \exp \mathfrak{a} = \exp(X+\mathfrak{a})$ because both sides give $f^{-1}f(X)$.

Suppose now $\mathfrak{n} = \mathfrak{n}_1 \supset \mathfrak{n}_2 \supset \dots$ is a filtration of the Lie algebra $\mathfrak{n}$ over k. Let $N = N_1 \supset N_2 \supset \dots$, $N_i = \exp \mathfrak{n}_i$, be the corresponding sequence of subgroups of N. This a filtration of the group N, by 2.5.6, and exp induces an isomorphism of the associated graded Lie algebras

$$\operatorname{gr} \mathfrak{n} \xrightarrow{\sim} \operatorname{gr} N$$

where $\operatorname{gr} \mathfrak{n}$ is regarded as a Lie algebra over $\mathbb{Z}$, by 2.5.5 and 2.5.6. The claim of the next proposition s that N_i, $i \in \mathbb{N}$, is the descending central series of the group N.

2.5.7 Proposition *Let $\mathfrak{n}$ be a nilpotent Lie algebra over a field k of characteristic zero. Let $\mathfrak{n} = \mathfrak{n}_1 \supset \mathfrak{n}_2 \supset \dots$ be the descending central series of $\mathfrak{n}$ and let $N = N_1 \supset N_2 \supset \dots$ be the descending central series of the corresponding Lie group. Then*

$$N_i = \exp \mathfrak{n}_i$$

for every $i \in \mathbb{N}$.

In particular, the exponential map induces an isomorphism of the associated graded Lie algebras

$$\operatorname{gr}\mathfrak{n} \xrightarrow{\sim} \operatorname{gr} N \quad .$$

We shall often identify $\operatorname{gr}\mathfrak{n}$ and $\operatorname{gr} N$ by this map and in particular regard $\operatorname{gr} N$ as a Lie algebra over k .

PROOF. Define $N_{(i)} = \exp \mathfrak{n}_i$. Since the descending central series N_i , $i \in \mathbb{N}$, is the fastest decreasing filtration of a group, we have $N_i \subset N_{(i)}$. If $\mathfrak{n}$ is m-step nilpotent, we have in particular $N_i = 1$ for $i > m$. Let $\operatorname{gr} N$ and $\operatorname{gr}' N$ be the associated graded Lie algebras of the filtrations N_i and $N_{(i)}$, $i \in \mathbb{N}$. The identity map of N induces a map of graded Lie algebras $\operatorname{gr} N \to \operatorname{gr}' N$. Obviously, $\operatorname{gr}_1 N \to \operatorname{gr}_1' N$ is surjective. The $\mathbb{Z}$-Lie algebras $\operatorname{gr} N$ and $\operatorname{gr}' N$ are generated by $\operatorname{gr}_1 N$ and $\operatorname{gr}_1' N$ resp., by 2.3.6 and 2.1.2 and the remark preceding proposition 2.5.7. So $\operatorname{gr} N \to \operatorname{gr}' N$ is surjective. Therefore the inclusion $N_i \to N_{(i)}$ induces a surjection $N_i \to N_{(i)}/N_{(i+1)}$. It follows now by induction over $j = m-i$, $i = 0,1, \ldots,$ that $N_i = N_{(i)}$ □

We have a number of corollaries of 2.5.7 concerning generators of N. With 2.3.8 we obtain

2.5.8 COROLLARY *Let* H *be a subgroup of* N . *We have* $H = N$, *if the natural map* $H \to \exp \mathfrak{n}^{ab}$ *is surjective* □

Let us introduce the following notations. If X is a subset of a Lie algebra $\mathfrak{g}$ over k , denote by $\langle X \rangle$ resp. $\langle X \rangle^{\mathfrak{g}}$ the smallest Lie subalgebra over k (the smallest ideal of $\mathfrak{g}$ over k) containing X .

2.5.9 COROLLARY *Let* $\{V_j \; ; \; j \in J\}$ *be a set of vector subspaces of* $\mathfrak{n}$ *over* k . *Then*

$$\exp \langle \bigcup_{j \in J} V_j \rangle = \langle \bigcup_{j \in J} \exp V_j \rangle \; .$$

PROOF. Denote $\mathfrak{a} = \langle \bigcup_{j\in J} V_j \rangle \subset \mathfrak{n}$, $A = \langle \bigcup_{j\in J} \exp V_j \rangle \subset \exp \mathfrak{n}$. We have $A \subset \exp \mathfrak{a}$. To prove equality we may assume that $\mathfrak{a} = \mathfrak{n}$. Let $\varphi : \mathfrak{n} \to \mathfrak{n}^{ab} =: \mathfrak{m}$ be the natural homomorphism of Lie algebras over k and let Φ be the corresponding Lie group homomorphism. By the preceding corollary we have to show that $\Phi(A) = \exp \mathfrak{m}$. Since $\Phi(A) = \langle \bigcup_{j\in J} \Phi \exp V_j \rangle = \langle \bigcup_{j\in J} \exp \varphi V_j \rangle$, we are reduced to the case that our Lie algebra, viz. $\mathfrak{m}$, is abelian, where our claim follows from the fact that a subgroup of a vector space generated by vector subspaces is a vector space □

2.5.10 COROLLARY *For every natural number* r *there is a natural number* c *such that for every Lie algebra* $\mathfrak{n}$ *over* $\mathbb{Q}$ *, which is nilpotent of step* $\leq r$ *, and every two elements* X,Y *of* $\mathfrak{n}$ *we have*

a) $\exp(X,Y) \in \langle \exp c^{-1} X , \exp c^{-1} Y \rangle$

and *b)* $\exp[X,Y] \in \langle \exp c^{-1} X , \exp c^{-1} Y \rangle'$.

PROOF. It suffices to prove the corollary for the free step r nilpotent Lie algebra $\mathfrak{n}$ over $\mathbb{Q}$ with two generators X and Y . Let $N = \exp \mathfrak{n}$. Choose a sequence n_i , $n \in \mathbb{N}$, of integers $\neq 0$, such that n_i divides n_{i+1} for every $i \in \mathbb{N}$ and every integer divides some n_i , e.g. $n_i = i!$. Then N is the union of the ascending sequence of subgroups $G^i = \langle \exp n_i^{-1} X , \exp n_i^{-1} Y \rangle$ by 2.5.9. Hence $\exp(X+Y) \in N = \bigcup G^i$, hence $\exp(X+Y) \in G^i$ for some $i \in \mathbb{N}$. Similarly $\exp[X,Y] \in \exp[\mathfrak{n},\mathfrak{n}] = (N,N) = \bigcup_i (G^i,G^i)$, here we use 2.5.7 for the first equation, hence $\exp[X,Y] \in (G^i,G^i)$ for some i □

2.5.11 COROLLARY *Let* $\{V_j , j \in J\}$ *be a set of vector subspaces of* $\mathfrak{n}$ *over* k *. Then*

$$\exp \langle \bigcup_{j\in J} V_j \rangle^{\mathfrak{n}} = \langle \bigcup_{j\in J} \exp V_j \rangle^{N} .$$

PROOF. Denote $\mathfrak{a} = \langle \bigcup_{j\in J} V_j \rangle^{\mathfrak{n}}$, $B = \langle \bigcup_{j\in J} \exp V_j \rangle^{N}$ and $\mathfrak{h} = \{X\in\mathfrak{n} ; \exp X \in B\}$. Since $\exp \mathfrak{a}$ is a normal subgroup of N , namely the kernel of the group homomorphism induced by the natural map $\mathfrak{n} \to \mathfrak{n}/\mathfrak{a}$, we have $\exp \mathfrak{a} \supset B$, hence $\mathfrak{a} \supset \mathfrak{b}$. On the other hand, if $g \in N$ then $\langle B,g\rangle' \subset B$, since the image of $\langle B,g\rangle$ in N/B is generated by g hence commutative. It follows that $\mathfrak{b}$ has the following property. If for some $x \in \mathfrak{n}$ we have $\mathbb{Q} \cdot x \subset \mathfrak{b}$, then $[x,y] \in \mathfrak{b}$ for every $y \in \mathfrak{n}$, by 2.5.10 b). Hence if V is a k-vector space contained in $\mathfrak{b}$ and $x_1, \dots, x_s$ are elements of $\mathfrak{n}$ then $\mathrm{ad}x_1 \circ \dots \circ \mathrm{ad}x_s V \subset \mathfrak{b}$. So B contains the group generated by $\bigcup \exp(\mathrm{ad}x_1 \circ \dots \circ \mathrm{ad}x_s V_j)$, $j \in J$, $x_1, \dots, x_s$ in $\mathfrak{n}$, which by 2.5.9 is the exp-image of $\langle \bigcup \mathrm{ad}x_1 \circ \dots \circ \mathrm{ad}x_s V_j$, $j \in J$, $x_1, \dots, x_s$ in $\mathfrak{n}\rangle$ which is $\mathfrak{a}$. Hence $\mathfrak{b} \supset \mathfrak{a}$ □

We now have the following generalization of 2.5.7.

2.5.12 PROPOSITION *Let* $\mathfrak{n}$ *be a nilpotent Lie algebra over a field* k *of characteristic zero. Let* $\mathfrak{a}$ *and* $\mathfrak{b}$ *be ideals of* $\mathfrak{n}$. *Then*

$$\exp[\mathfrak{a},\mathfrak{b}] = (\exp \mathfrak{a}\ , \exp \mathfrak{b}) .$$

PROOF. $[\mathfrak{a},\mathfrak{b}]$ is an ideal of $\mathfrak{n}$ (see 2.1). Let $\mathfrak{m} = \mathfrak{n}/[\mathfrak{a},\mathfrak{b}]$. In $\exp \mathfrak{m}$ we have for x in the image of $\exp \mathfrak{a}$ and y in the image of $\exp \mathfrak{b}$ the equation $(x,y) = e$ by 2.5.6. Therefore $(\exp \mathfrak{a}, \exp \mathfrak{b}) \subset \exp[\mathfrak{a},\mathfrak{b}]$. To see the converse inclusion note that $[\mathfrak{a},\mathfrak{b}]$ is the smallest Lie algebra containing all the vector spaces $k[x,y]$, $x \in \mathfrak{a}$, $y \in \mathfrak{b}$. So by 2.5.9 it suffices to prove that $\exp k[x,y] \in (\exp \mathfrak{a}, \exp \mathfrak{b})$ for every $x \in \mathfrak{a}$, $y \in \mathfrak{b}$, i.e. that $\exp[x,y] \in (\exp \mathfrak{a}, \exp \mathfrak{b})$ for every $x \in \mathfrak{a}$, $y \in \mathfrak{b}$. But $\exp[x,y] \in \langle \exp c^{-1}x, \exp c^{-1}y\rangle'$ for some $c \in \mathbb{N}$ by 2.5.10 b), and the group $\langle \exp c^{-1}x, \exp c^{-1}y\rangle$ is generated by two elements which commute modulo $(\exp \mathfrak{a}, \exp \mathfrak{b})$, so $\langle \exp c^{-1}x, \exp c^{-1}y\rangle' \subset (\exp \mathfrak{a}, \exp \mathfrak{b})$ □

We shall need the following generalization of 2.5.10 a) later on, see 5.3 and 5.4. The proposition is a comparison result comparing the Lie subring (= Lie subalgebra over $\mathbb{Z}$) and the subgroup generated by corresponding sets.

2.5.13 PROPOSITION *For every natural number* r *there is a natural number* c *with the following property. Given a nilpotent Lie algebra* $\mathfrak{n}$ *over* $\mathbb{Q}$ *of step* $\leq r$ *and a subset* X *of* $\mathfrak{n}$. *Let* $\mathfrak{m}$ *be the Lie subring generated by* X *and let* A *be the subgroup of* $N = \exp \mathfrak{n}$ *generated by* $\exp X$. *Then there is a Lie subring* $\mathfrak{h}$ *of* $\mathfrak{n}$ *which is the logarithm of a subgroup of* N, *such that*

$$\mathfrak{h} \supset \mathfrak{m} \cup \log A \supset \mathfrak{m} \cap \log A \supset c\mathfrak{h} .$$

In particular

$$c\mathfrak{m} \subset \log A \subset c^{-1}\mathfrak{m} .$$

PROOF. We may assume that $\mathfrak{n}$ is the free step r nilpotent Lie algebra over $\mathbb{Q}$ on the set X of generators. We have to make sure that our constant c depends only on r, not on X. Let $\mathfrak{m}$ be the Lie subring of $\mathfrak{n}$ generated by X. Let $\mathfrak{m} = \mathfrak{m}_1 \supset \mathfrak{m}_2 \supset \ldots$ and $\mathfrak{n} = \mathfrak{n}_1 \supset \mathfrak{n}_2 \supset \ldots$ be the descending central series of $\mathfrak{m}$ and $\mathfrak{n}$ resp. For $j \in \mathbb{N}$ let $X^j \subset \mathfrak{n}$ be the set of Lie monomials of degree j with entries from X. Denote by $\mathfrak{n}^j$ the vector subspace over $\mathbb{Q}$ spanned by X^j and by $\mathfrak{m}^j$ the $\mathbb{Z}$-submodule of $\mathfrak{n}$ spanned by X^j. These define gradations of the Lie algebras $\mathfrak{n}$ and $\mathfrak{m}$, since $\mathfrak{n} = \oplus\, \mathfrak{n}^j$ by [46] LA IV Th. 4.2. It follows that $\mathfrak{m}^j = \mathfrak{m} \cap \mathfrak{n}^j$, $\mathfrak{n}_i = \bigoplus_{j\geq i} \mathfrak{n}^j$ and $\mathfrak{m}_i = \bigoplus_{j\geq i} \mathfrak{m}^j = \mathfrak{m} \cap \mathfrak{n}_i$. In particular $\operatorname{gr} \mathfrak{m} \simeq \mathfrak{m}$. We shall need the following consequence. As a Lie ring $\operatorname{gr} \mathfrak{m}$ is generated by the image of X.

I claim that there is a sequence of integers $m_j \neq 0$, $j = 1,\ldots r$, with $m_1 = 1$ and m_j dividing m_{j+1}, such that $\mathfrak{h} = \Sigma\, m_k^{-1} \mathfrak{m}_k$ is the logarithm of a subgroup of N. The point is that these numbers m_j do not depend on X, only on r. To see this, let n_i be an integer $\neq 0$, such that n_i times the homogeneous part of the Campbell-Hausdorff formula of degree i has integer coefficients. Set $n_1 = 1$. Let $X = \Sigma X_k \in \mathfrak{h}$, $Y = \Sigma Y_k \in \mathfrak{h}$, X_k and Y_k in $m_k^{-1}\mathfrak{m}_k$. Then $X \circ Y$ is a sum of terms of the following form: coefficient times iterated Lie brackets with s entries $X_{i_1}, \dots X_{i_s}$ and t entries $Y_{j_1}, \dots Y_{j_t}$. This term is in $m_{i_1}^{-1} \dots m_{i_s}^{-1} m_{j_1}^{-1} \dots m_{j_t}^{-1} n_{s+t}^{-1} \cdot m_{i_1+\dots+i_s+j_1+\dots+j_t}$. Hence if the m_j's are so determined that $m_{i_1} \dots m_{i_s} m_{j_1} \dots m_{j_t} \cdot n_{s+t}$ divides $m_{i_1+\dots+i_s+j_1+\dots+j_t}$ for every set $i_1, \dots i_s, j_1, \dots j_t$ of natural numbers - which is obviously possible inductively - , then the $\mathbb{Z}$-module $\mathfrak{h}$ is the logarithm of a subgroup of N and is a Lie ring, since $m_i \cdot m_j$ divides m_{i+j}. Now $\mathfrak{h} \supset \mathfrak{m} \cup \log A$. On the other hand $m_r \cdot \mathfrak{h} \subset \mathfrak{m}$.

It remains to prove that there is an integer $c \neq 0$ such that for every element $Y \in \mathfrak{h}$ we have $cY \in \log A$, or equivalently that $x \in B = \exp \mathfrak{h}$ implies $x^c \in A$. Look at the filtrations of $A, B, \mathfrak{h}$ and $\mathfrak{m}$ defined by intersecting with the descending central series $N_i = \exp \mathfrak{n}_i$ and $\mathfrak{n}_i$ resp. Then we have injective homomorphisms of the corresponding graded Lie algebras

$$\operatorname{gr} \mathfrak{n} \supset \operatorname{gr} \mathfrak{h} \cong \operatorname{gr} B \supset A \supset \operatorname{gr} \mathfrak{m} ,$$

where the associated graded Lie algebras of subgroups of N are embeded into $\operatorname{gr} \mathfrak{n}$ via the logarithm using 2.5.7. The image of $\operatorname{gr} \mathfrak{m}$ is contained in $\operatorname{gr} A$, since both are containd in $\operatorname{gr} \mathfrak{n}$ and $\operatorname{gr} \mathfrak{m}$ is generated as a Lie ring by the image of X in $\operatorname{gr}_1 \mathfrak{m}$, which is contained in $\operatorname{gr}_1 A$. We have proved that $m_r \mathfrak{h} \subset \mathfrak{m}$. This implies $m_r(\operatorname{gr} B) \subset \operatorname{gr} A$, hence $x^{(m_r)^i} \in A/A_i$ for $x \in B/B_i$ by induction □

2.5.14 REMARK Let $\mathfrak{n}$ be a nilpotent Lie algebra over a field k of characteristic zero. If $\mathfrak{a}$ is a Lie subring of $\mathfrak{n}$, then $\exp \mathfrak{a}$ need not be a subgroup of $\exp \mathfrak{n}$, in general, since the Campbell-Hausdorff formula has denominators. Neither conversely, if Γ is a subgroup of $\exp \mathfrak{n}$, then $\log \Gamma$ is not a Lie subring of $\mathfrak{n}$ is general. E.g. let $\mathfrak{n}$ be the free step 2 nilpotent Lie algebra over $\mathbb{Q}$ on two generators X and Y - a rational form of the Heisenberg algebra. Then for

$\Gamma = \langle \exp X, \exp Y \rangle$ the logarithm $\log \Gamma = \{xX + yY + (z - \frac{1}{2}xy)[X,Y] ; x,y,z \text{ in } \mathbb{Z}\}$ is not a Lie ring, since X,Y are in $\log \Gamma$ but $X + Y$ is not in $\log \Gamma$.

2.6 Nilpotent Lie groups over p-adic fields

In this section we assume $\mathfrak{n}$ to be a nilpotent Lie algebra of finite dimension over a p-adic field K. Then $N = \exp \mathfrak{n}$ is a topological group. We show that every compact subset of N is contained in a compact subgroup of N. This implies the announced converse of 1.3.1. A local field K will be called p-adic, if it has characteristic zero and its natural valuation (see 1.2), the module mod_K is non-archimedian. Then K is a finite extension of a field $\mathbb{Q}_p$ of p-adic numbers for some prime number p. Let $\mathfrak{o}$ be the valuation ring $\mathfrak{o} = \{x \in K ; \operatorname{mod}_K(x) \le 1\}$ of K and let $\mathfrak{m}$ be its maximal ideal $\mathfrak{m} = \{x \in K; \operatorname{mod}_K(x) < 1\}$. Then p is the characteristic of $\mathfrak{o}/\mathfrak{m}$.

Let $\mathfrak{n}$ be a finite-dimensional nilpotent Lie algebra over a p-adic field K and let $N = \exp \mathfrak{n}$ be the corresponding Lie group, defined by Campbell-Hausdorff multiplication (see 2.5). $\mathfrak{n}$ has a unique topology for which it is a topological vector space over K. For the topology on N, for which $\exp : \mathfrak{n} \to N$ is a homeomorphism, N is a topological group, since the Campbell-Hausdorff formula gives a polynomial map $\mathfrak{n} \times \mathfrak{n} \to \mathfrak{n}$, $(X,Y) \to X \circ (-Y)$.

Recall that we call a Lie algebra over $\mathbb{Z}$ a Lie ring.

2.6.1 Lemma *The open compact Lie subrings which are logarithms of subgroups of* N *form a neighborhood base of zero in* $\mathfrak{n}$.

Proof. Let $X_1, \ldots X_m$ be a basis of the Lie algebra $\mathfrak{n}$ regarded as vector space over K. The structure constants $c_{ij}^k \in K$ with respect to the basis $X_1, \ldots X_m$ are defined by the equations

$$[X_i, X_j] = \sum_k c^k_{ij} X_k \ .$$

Multiplying every X_j by an element $d^{-1} \in K$ such that $\text{mod}_K(d) = \max \{\text{mod}_K(c^k_{ij}) \ ; \ i,j,k \in \{1, \ldots m\}\}$ we obtain a basis $Y_1, \ldots Y_m$ of $\mathfrak{n}$ with all structure constants in the valuation ring o . Then the set of open compact Lie subrings $a \cdot \sum_j o\, Y_j$, $a \in K^*$, forms a neigborhood base of zero in $\mathfrak{n}$.

Now let U be a neighborhood of zero in $\mathfrak{n}$, let c be as in 2.5.13 for $\mathfrak{n}$. Apply 2.5.13 for X an open compact Lie subring $\mathfrak{m}$ of cU to obtain a Lie subring $\mathfrak{h} \subset c^{-1}\mathfrak{m} = c^{-1}X \subset U$ which contains $\mathfrak{m} = X$ and is the logarithm of a subgroup of N . The additive group $\mathfrak{h}$ contains the neighborhood $\mathfrak{m}$ of zero, hence is open, hence closed, hence compact, since contained in the compact set $c^{-1} \cdot \mathfrak{m}$ □

We shall need the following lemma later on.

2.6.2 LEMMA a) *A subgroup* A *of* N *is open iff its image in* N^{ab}
is open.

b) *If* A *is open in* N , *then the i-th term* A_i *of the descending central series of* A *is open in* N_i *for every* $i \in \mathbb{N}$.

PROOF. a) $N \to N^{ab}$ is an open homomorphism, therefore the condition is necessary. To prove sufficiency, we may assume $K = \mathbb{Q}_p$, because restricting scalars to $\mathbb{Q}_p$ does not change the topological group N . Again, let $\text{gr}\, N$ be the associated graded Lie algebra of the descending central series $N = N \supset N_2 \ldots$ of N . The Lie ring $\text{gr}\, N$ has a topology induced from N , for which it is a topological Lie ring and the exponential map induces an isomorphism $\text{gr}\, \mathfrak{n} \xrightarrow{\sim} \text{gr}\, N$ of topological Lie rings, by 2.5.7 and the definition of the topology on N . So we may regard $\text{gr}\, N$ as a topological Lie algebra over K . Define a filtration on A by $A_i = A \cap N_i$ and let $\text{gr}\, A$ be the associated graded Lie ring, $\text{gr}\, A \rightarrowtail \text{gr}\, N$. For every $k \in \mathbb{N}$ the subgroup $\text{gr}_k A$ of $\text{gr}_k N$

spans the K-vector space $gr_k N$, since $gr_1 A$ does by hypothesis and, inductively, $gr_k N$ is spanned by the image of the Lie bracket $gr_1 N \times gr_{k-1} N \to gr_k N$, which is spanned by the image of $gr_1 A \times gr_{k-1} A$, a subset of $gr_k A$. So $\log(A)$ contains a basis of the K-vector space $\mathfrak{n}_k = \log(N_k)$ by descending induction. Let B be a basis of $\mathfrak{n}$ contained in $\log(A)$. Then for every $X \in B$ we have $\mathbb{Z}_p \cdot X \subset \log(A)$, since A is an open, hence closed subgroup of N . Let c be as in 2.5.10. Then the open subset $c^r \cdot \sum_{X \in B} \mathbb{Z}_p X$ of $\mathfrak{n}$ is containd in $\log(A)$, with $r = \dim \mathfrak{n}$, by 2.5.10 a). So the group A contains a neighborhood of e , hence is open.

b) Again, assume $K = \mathbb{Q}_p$. Let B be a basis of $\mathfrak{n}$ such that $\mathbb{Z}_p X \subset \log(A)$ for every $X \in B$. Assume by induction A_i is open in N_i for some $i \in \mathbb{N}$. Then there is a basis B_i of $\mathfrak{n}_i$ such that $\mathbb{Z}_p X \subset \log(A_i)$ for every $X \in B_i$. With c as in 2.5.10 we have $c^2[X,Y] \in \log\langle \exp X , \exp Y\rangle' \subset \log A_{i+1}$ for $X \in B$, $Y \in B_i$, hence $c^n \cdot \sum_{X \in B, Y \in B_i} \mathbb{Z}_p [X,Y] \subset \log A_{i+1}$ with $n = 2 + \dim \mathfrak{n} + \dim \mathfrak{n}_i$ by 2.5.10, so A_{i+1} is open in $\mathfrak{n}_{i+1}$ □

We give three proofs of the following proposition, which will imply the converse of 1.3.1 we were looking for.

2.6.3 PROPOSITION *Every compact subset of* N *is contained in a compact subgroup of* N .

1ST PROOF. Let U be an open compact subgroup of N . The image V of U in N^{ab} is open and compact, so N^{ab} is the isolator of V . Actually if p is the characteristic of the residue field o/m of K , then N^{ab} is the p-isolator of V , see 2.4.3. Hence N is the p-isolator of U , by 2.4.3. Now let C be a compact subset of N . We have to prove that $\langle C\rangle^-$ is compact. We may assume that $U \subset C$. Then C is containd in a finite set of U-cosets. So $\langle C\rangle$ is generated by U and finite subset of the p-isolator of U , hence U is of finite index in $\langle C \rangle$ by 2.4.4, so $\langle C \rangle$ is compact.

<u>2ND PROOF,</u> by induction on $\dim \mathfrak{n}$. If $\dim \mathfrak{n} = 1$, N is isomorphic to the additive group of K. If C is a compact subset of K, the module mod_K is bounded on C, by R say. Then $\{x \in K ; \mathrm{mod}_K(x) \leq R\}$ is a compact additive subgroup of K.

Suppose $\dim \mathfrak{n} > 1$. Let $\mathfrak{m}$ be a codimension one ideal of $\mathfrak{n}$, i.e. the inverse image of a codimension one subspace of $\mathfrak{n}^{ab}$ under $\mathfrak{n} \to \mathfrak{n}^{ab}$. Let X be an element of $\mathfrak{n}$ not in $\mathfrak{m}$. Define $M = \exp \mathfrak{m}$ and $J = \exp(K \cdot X)$. Multiplication defines a homeomorphism $M \times J \to N$. The inverse mapping can be obtained as follows. The natural map $\mathfrak{n} \to \mathfrak{n}/\mathfrak{m}$ restricts to an isomorphism of Lie algebras $\varphi : K \cdot X \to \mathfrak{n}/\mathfrak{m}$. From the homomorphism $s : N \to J$ of Lie groups induced by $\mathfrak{n} \to \mathfrak{n}/\mathfrak{m} \xrightarrow{\varphi^{-1}} K \cdot X$ we obtain the inverse map $N \to M \times J$, $u \longrightarrow (u \cdot s(u)^{-1}, s(u))$.

Let C be a compact subset of N. There are compact subgroups C_1 of M and C_2 of J such that $C \subset C_1 \cdot C_2$ by the inductive hypothesis. The inner automorphisms of N define a continuous action of N on itself. So $C_3 = \{y x y^{-1} ; y \in C_2, x \in C_1\}$ is a compact C_2-invariant subset of M. The closure C_4 of the subgroup of M generated by C_3 is compact, by our inductive hypothesis. C_2 normalizes C_4, so $C_4 \cdot C_2$ is a compact group containing C.

<u>3RD PROOF.</u> By a theorem about algebraic groups (cf. [26]) N may be regarded as the set of K-points of a unipotent linear algebraic group defined over K, in particular as a closed subgroup of the group U of upper triangular $n \times n$ - matrices with entries from K and ones on the diagonal. So to prove our theorem we may assume that $U = N$. For $R \in \mathbb{R}$ define U_R as the set of matrices $(a_{ij}) \in U$ satisfying

$$\mathrm{mod}_K(a_{ij}) \leq R^{j-i}$$

for every entry a_{ij}. Now U_R is a compact subgroup of U and every compact subset of U is contained in U_R for some $R > 1$ □

2.6.4 THEOREM *Let* $\mathfrak{n}$ *be a nilpotent finite-dimensional Lie algebra over a p-adic field* K *. Let* $N = \exp \mathfrak{n}$ *be the corresponding Lie group. Let* β *be an automorphism of* $\mathfrak{n}$ *and let* α *be the corresponding automorphism of* N *. Let the infinite cyclic group* $\langle t \rangle$ *act on* N *by sending* t *to* α *. The split extension* $\langle t \rangle \ltimes N$ *has a compact presentation iff* β *or* β^{-1} *acts contracting on* $\mathfrak{n}$ *.*

PROOF. Sufficiency was proved in 1.3.1. In 1.3.2 it was shown, that if $\langle t \rangle \ltimes N$ has a compact presentation then there is an open subgroup B of N having a compact set of generators such that 1) $\alpha^{-1}(B) \subset B$ and $\bigcup_{n \in \mathbb{N}} \alpha^n(B) = N$, or 2) $\alpha(B) \subset B$ and $\bigcup_{n \in \mathbb{N}} \alpha^{-n}(B) = N$. Since open subgroups of topological groups are closed, B is compact by 2.6.3. By considering α^{-1} instead of α we may assume that we are in the second case. So there is an open compact neighborhood U of $\mathfrak{n}$ such that $\beta(U) \subset U$ and $\bigcup_{n \in \mathbb{N}} \beta^{-n}(U) = N$. I show that β is contracting on $\mathfrak{n}$. Let V be a neighborhood of 0 in $\mathfrak{n}$ and let C be a compact subset of $\mathfrak{n}$. We have to show that there is a number $n \in \mathbb{N}$ such that $\beta^n C \subset V$. There is a non zero element $k \in K$ such that $kU \subset V$. The ascending sequence $\beta^{-n}(U)$, $n \in \mathbb{N}$, covers the compact set $k^{-1} \cdot C$, so $k^{-1}C \subset \beta^{-n}(U)$ for some $n \in \mathbb{N}$, hence by linearity $\beta^n C \subset kU \subset V$ □

III. A NECESSARY CONDITION FOR COMPACT PRESENTABILITY

Let Q be a finitely generated free abelian group of rank n . Let K be a p-adic field, i.e. a local field of characteristic zero with non-archimedian natural valuation, or equivalently a finite extension field of some $\mathbb{Q}_p$. Let $\mathfrak{n}$ be a finite dimensional nilpotent Lie algebra over K . Suppose a homomorphism from Q to the group of Lie algebra automorphisms of $\mathfrak{n}$ is given. It defines an action of Q on the corresponding Lie group $N = \exp \mathfrak{n}$. In this and the following chapters we deal with the following

PROBLEM. *Under which conditions is the semidirect product* $Q \ltimes N$ *compactly pesentable?*

The case rank $Q = 1$ is a solved by theorem 2.6.4. In this chapter we give a necessary condition, called condition 1), for arbitrary rank of Q (theorem 3.2.3).

The contents of this chapter are as follows. In section 3.1 we establish notations which will be in force for the following chapters. In section 3.2 we prove a necessary and sufficient condition for compact generation of $Q \ltimes N$ due to Borel and Tits [22 Theorem 13.4], theorem 3.2.2. We give the necessary condition mentioned above for compact presentability. This is analogous to a theorem of Bieri and Strebel [13, 14]. The analogy is explained in section 3.4. We show that the necessary condition of 3.2.3 is not sufficient by giving an example in section 3.3. In the next chapters we shall analyze the consequences of the necessary condition 1). Furthermore we restate the solution of the problem for the case rank $Q = 1$ in our present terminology, theorem 3.2.4.

3.1 WEIGHTS

Let Q be a finitely generated free abelian group. Let K be a p-adic field. Let ρ be a representation of Q on a K-vector space V. For every $\lambda \in \mathrm{Hom}(Q,K^*)$ the *weight space* of λ is

$$V^\lambda = \{v \in V \; ; \; \rho(t)v = \lambda(t)v \quad \text{for every} \quad t \in Q\} .$$

The *set of weights* $W(\rho) = W(Q,V) = W(V)$ of the Q-module V is

$$W(V) = \{\lambda \in \mathrm{Hom}(Q,K^*) \; ; \; V^\lambda \neq 0\} .$$

The set of Q-submodules V^λ, $\lambda \in W(V)$, is linearly independent, i.e. a sum Σv_λ, $v_\lambda \in V^\lambda$, is zero only if all the v_λ are zero. Therefore the K[Q]-module V is a sum of one-dimensional submodules iff $V = \oplus V^\lambda$, $\lambda \in W(V)$.

Let Q act on the Lie algebra $\mathfrak{n}$ over K by automorphisms, i.e. suppose given a homomorphism ρ from Q to the group of K-Lie algebra automorphisms of $\mathfrak{n}$. Then

(3.1.1) $$[\mathfrak{n}^\lambda, \mathfrak{n}^\mu] \subset \mathfrak{n}^{\lambda+\mu}$$

for every pair of element λ,μ in $\mathrm{Hom}(Q,K^*)$, where the pointwise multiplication in the group $\mathrm{Hom}(Q,K^*)$ is written additively.

3.1.2 LEMMA *Let* Q *act on* $\mathfrak{n}$ *by automorphisms. Suppose* $\mathfrak{n}$ *is a sum of one-dimensional* K[Q]*-submodules. Then every weight in* $W(\mathfrak{n})$ *is a non-trivial linear combination of weights in* $W(\mathfrak{n}^{ab})$ *with non-negative integer coefficients.*

PROOF. Let W' be the set of non-trivial linear combination of weights in $W(\mathfrak{n}^{ab})$ with non-negative integer coefficients. W' is a subsemigroup of $\mathrm{Hom}(Q,K^*)$ containing $W(\mathfrak{n}^{ab})$. So $\mathfrak{h} := \oplus\, \mathfrak{n}^{\mu}$, $\mu \in W'$, is a Lie subalgebra of $\mathfrak{n}$, by 3.1.1. Our claim $\mathfrak{h} = \mathfrak{n}$ will follow from 2.1.3 when we have shown that the natural map $\pi : \mathfrak{n} \to \mathfrak{n}^{ab}$ maps $\mathfrak{h}$ onto $\mathfrak{n}^{ab}$. If $\lambda \notin W(\mathfrak{n}^{ab})$ we have $\pi(\mathfrak{n}^{\lambda}) \subset \mathfrak{n}^{ab,\lambda} = 0$, since π is a map of $K[Q]$-modules. So $\mathfrak{n}^{ab} = \pi(\mathfrak{n}) = \pi(\sum_{\lambda \in W(\mathfrak{n})} \mathfrak{n}^{\lambda}) = \pi(\sum_{\lambda \in W(\mathfrak{n}^{ab})} \mathfrak{n}^{\lambda}) \subset \pi(\mathfrak{h})$ □

From now on we make the hypothesis

3.1.3 HYPOTHESIS $\mathfrak{n}$ *is the (direct) sum of on-dimensional* $K[Q]$*-submodules.*

3.1.4 Let mod_K be the natural valuation of K (see paragraph preceding 1.2.3). Define an exponential valuation ν of K by

$$\nu = -\log \circ \mathrm{mod}_K$$

So we have

$$\nu(x \cdot y) = \nu(x) + \nu(y)$$

$$\nu(x+y) \geq \inf(\nu(x), \nu(y)) \quad .$$

Composing with ν gives a homomorphism

$$\nu_* : \mathrm{Hom}(Q,K^*) \to \mathrm{Hom}(Q,\mathbb{R})$$

$$\nu_*(\mu) = \nu \circ \mu \quad .$$

Our field K is a finite extension of some $\mathbb{Q}_p$, so the image of mod_K is a discrete subgroup of $\mathbb{R}^*_+$ and hence the image of ν_* is a lattice in the vector space $\mathrm{Hom}(Q,\mathbb{R})$.

3.1.5 Instead of weights $\mu \in \mathrm{Hom}(Q,K^*)$ we shall mostly consider their *values* $\nu_*(\mu) \in \mathrm{Hom}(Q,\mathbb{R})$. So for every $K[Q]$-module V and $\alpha \in \mathrm{Hom}(Q,\mathbb{R})$ we set

$$V^{\alpha} = \bigoplus_{\nu_*(\mu)=\alpha} V^{\mu} \quad .$$

Note that $t \in Q$ acts contracting on V^{α} iff $\alpha(t) < 0$, by 1.2.3.

3.1.6 Let π be a Lie algebra over K and let Q act on π by automorphisms. The most important geometric invariant we associate to this action will be the following

$$R = \nu_* W(\pi^{ab}) \quad .$$

This is closely related to the Bieri-Strebel invariant of the $K[Q]$-module π^{ab} (see 3.4 and 7.3). Recall that we assume that π is a sum of one dimensional $K[Q]$-submodules.

3.2 VALUES OF WEIGHTS

In this section we use the set R just defined in order to state and prove some basic results about compact generation and compact presentability of groups of the form $Q \ltimes \exp \pi$. We give a necessary and sufficient condition for compact generation (theorem 3.2.2), a necessary condition for compact presentability in general (theorem 3.2.3) and a necessary and sufficient condition for compact presentability in case Q is infinite cyclic (theorem 3.2.4, rephrasing 2.6.4). We shall see in the next section that the set R does not decide the question of compact presentability, in general.

We shall need the following terminology.

3.2.1 Let T be a subset of a real vector space. A linear combination of T is a sum $\Sigma\lambda_t \cdot t$, $t \in T$, $\lambda_t \in \mathbb{R}$, such that all the λ_t are zero except for a finite subset of T . The linear combination $\Sigma\lambda_t \cdot t$, $t \in T$, is called a *positive linear combination*, if $\lambda_t \geq 0$ for every $t \in T$. The set T is called *positively independent* if a positive linear combination $\Sigma\lambda_t \cdot t$ of T is zero iff $\lambda_t = 0$ for every $t \in T$. So T is positively independent iff the zero vector is not in the convex hull of T . A set T consisting of two elements is positively independent iff both elements are not zero and there is no straight line containing both on opposite sides of zero iff for both elements $t \in T$ the ray $\{rt \mid r \le 0\}$ does not intersect T. We use the notations and hypotheses of 3.1. Recall that the set R of 3.1.6 is a subset of the real vector space $\mathrm{Hom}(Q,\mathbb{R})$ of dimension $n = \mathrm{rank}\, Q$.

3.2.2 THEOREM *$Q \ltimes N$ is compactly generated iff $0 \notin R$.*

This is basically a special case of a theorem of Borel and Tits [22, Theorem 13.4]

3.2.3 THEOREM *If $Q \ltimes N$ has a compact presentation, then condition 1) holds: Any two elements of R are positively independent.*

In a different terminology (see 0.2.15, 7.3.10) condition 1) can be restated as follows: The presentation of Q on $\mathfrak{n}^{ab}$ is tame. This is analogous to a result of Bieri and Strebel [13,14], cf. 3.4.

3.2.4 THEOREM *Suppose Q is infinite cyclic with generator t. Then $Q \ltimes N$ has a compact presentation iff $\lambda(t) > 0$ for every $\lambda \in R$ or $\lambda(t) < 0$ for every $\lambda \in R$.*

This is just a restatement of 2.6.4, as the following proof shows.

PROOF OF 3.2.4 $\langle t \rangle \ltimes N$ has a compact presentation iff t or t^{-1} acts contracting on $\mathfrak{n}$, by 2.6.4, iff $\lambda(t) > 0$ for every $\lambda \in \nu_* W(\mathfrak{n})$ or $\lambda(t) < 0$ for every $\lambda \in \nu_* W(\mathfrak{n})$, by 1.2.3, iff $\lambda(t) > 0$ for every $\lambda \in R$ or $\lambda(t) < 0$ for every $\lambda \in R$, by 3.1.2 □

PROOF OF 3.2.2 Suppose $Q \ltimes N$ has a compact set of generators. Then so does the factor group $Q \ltimes N^{ab} = Q \ltimes \exp \mathfrak{n}^{ab}$. Hence so does the group $Q \ltimes \exp \mathfrak{n}^{ab,0}$, where $\mathfrak{n}^{ab,0} = \bigoplus_{\nu_*(\lambda)=0} \mathfrak{n}^{ab} = \mathfrak{n}^{ab} / \bigoplus_{\nu_*(\lambda)\neq 0} \mathfrak{n}^{ab,\lambda}$. Here $\exp : \mathfrak{n}^{ab,0} \to \exp \mathfrak{n}^{ab,0}$ is an isomorphism of abelian topological groups. Let X_i, $i = 1,\dots,m$, be a basis of $\mathfrak{n}^{ab,0}$ consisting of weight vectors. Let $\mathcal{O} = \{x \in K \,;\, \mathrm{mod}_K(x) \leq 1\}$ be the valuation ring of K and let $\mathcal{O}^*$ be its group of units. We have $\rho(t)X_i \in \mathcal{O}^*X_i$ for every $t \in Q$, since $\nu_*(\lambda) = 0$, so $\rho(t)X_i = \lambda(t)X_i \in \mathcal{O}^* X_i$, hence $Q \ltimes \sum_i \mathcal{O} X_i$ is a subgroup of $Q \ltimes \mathfrak{n}^{ab,0}$. For every compact subset C of $Q \ltimes \mathfrak{n}^{ab,0}$ there is a basis X_i, $i = 1,\dots,m$, such that $C \subset Q \ltimes \Sigma\mathcal{O}\cdot X_i$. Therefore $Q \ltimes N$ is compactly generated only if $\dim \mathfrak{n}^{ab,0} = 0$.

Conversely, suppose $0 \notin R$. Let T be a finite set of generators of Q and let U be a compact neighborhood of 1 in N. I claim that $T \cup U$ generates $Q \ltimes N$. Clearly, the subgroup H of $Q \ltimes N$ generated by $T \cup U$ contains Q and $H \cap N$ is a Q-invariant subgroup of N containing U. By proposition 2.3.8 it suffices to prove that $H \cap N \longrightarrow N^{ab}$ is surjective. Let us identify $\mathfrak{n}^{ab}$ with N^{ab}. The image H_1 of $H \cap N$ in $\mathfrak{n}^{ab}$ is a Q-invariant subgroup of $\mathfrak{n}^{ab,\lambda}$. So $H_1 = \mathfrak{n}^{ab}$ □

<u>PROOF OF 3.2.3</u> The theorem will be proved by first reducing it to the case that R is contained in a one dimensional vector subspace of $\mathrm{Hom}(Q,\mathbb{R})$, then reducing to the case that Q has rank one and finally applying 3.2.4.

Let $\lambda \in W(\mathfrak{n})$ be a weight with $\nu_*(\lambda) \neq 0$. Then there is an element $t \in Q$ that induces a contracting linear isomorphism of $\mathfrak{n}^\lambda$. Therefore $\exp \mathfrak{n}^\lambda$ is the smallest Q-invariant subset of $\exp \mathfrak{n}^\lambda$ containing a neighborhood of the identity in the relative topology of $\exp \mathfrak{n}^\lambda$.

Let S be a subset of $\nu_* W(\mathfrak{n})$ such that $\lambda \neq 0$ for every $\lambda \in S$. Let $\mathfrak{a}$ be the smallest ideal of $\mathfrak{n}$ containing $\cup\, \mathfrak{n}^\lambda$, $\lambda \in S$. Then $A := \exp \mathfrak{a}$ is a normal subgroup of $Q \cdot N := Q \ltimes N$, compactly generated as a normal subgroup of $Q \cdot N$, by the preceding remark and 2.5.11. Hence, if $Q \cdot N$ has a compact presentation, so does $Q \cdot N/A$, by proposition 1.1.3 c).

Suppose for the rest of the proof that $Q \cdot N$ has a compact presentation and that there are two elements λ_+, λ_- of $R = \nu_* W(\mathfrak{n}^{ab})$ such that λ_+ and λ_- are positively dependent. Note that $\lambda_+ \neq 0$ and $\lambda_- \neq 0$ by 3.2.2. So there is a negative real number r such that $\lambda_- = r \cdot \lambda_+$. Let ℓ be the one dimensional real vector subspace of $\mathrm{Hom}(Q,\mathbb{R})$ containing λ_+ and λ_-, a line.

We first show that we may assume that $\nu_* W(\mathfrak{n}) \subset \ell$. To see this let S be the set of $\mu \in \nu_* W(\mathfrak{n})$ such that $\mu \notin \ell$. Let A be as above, i.e. $A = \exp \mathfrak{a}$, where $\mathfrak{a}$ is the smallest ideal of $\mathfrak{n}$ containing $\cup \mathfrak{n}^{\mu}$, $\mu \in S$. Let $\mathfrak{m} = \mathfrak{n}/\mathfrak{a}$, $M = \exp \mathfrak{m} = N/A$. Then $Q \cdot M = Q \ltimes M$ has a compact presentation, by the remarks above. Obviously $\mathfrak{a} \subset \mathfrak{b} := \sum_{\mu \in S} \mathfrak{n}^{\mu} + \mathfrak{n}'$, $\mathfrak{m}^{ab} = \mathfrak{n}/\mathfrak{b}$, so $\lambda_+, \lambda_- \in \nu_* W(\mathfrak{m}^{ab})$, and $\nu_* W(\mathfrak{m}) \subset \ell$.

So we assume form now on that $R = \nu_* W(\mathfrak{n}^{ab})$ is containd in the line $\ell \subset \mathrm{Hom}(Q,\mathbb{R})$. Then $\nu_* W(\mathfrak{n}) \subset \ell$ by 3.1.2. We shall reduce the proof to the case that Q is infinite cyclic. Note that any two non-zero elements λ of ℓ are multiples of each other, so they have the same kernel, say Q'. The image of every homomorphism $\lambda \neq 0$ in $\nu_* \mathrm{Hom}(Q,K^*)$ is infinite cyclic, since the image of ν is infinite cyclic. So Q/Q' is infinite cyclic. Let t be an element of Q whose residue class modulo Q' generates Q/Q'. We shall show that $\langle t \rangle \ltimes N$ has a compact presentation. Since $\lambda_+(t) \neq 0$ and $\lambda_-(t) \neq 0$ and $\lambda_- = r \cdot \lambda_+$, $r < 0$, we are then reduced to the case that Q is infinite cyclic.

For every $\lambda = \nu_*(\alpha) \in \ell$, $\alpha \in W(\mathfrak{n})$, every element $t' \in Q' \subset \ker \lambda$ acts on $\mathfrak{n}^{\alpha}$ by multiplication with a unit $\alpha(t') \in o^*$. Hence if we denote the representation of Q on $\mathfrak{n}$ by $\rho : Q \to \mathrm{Aut}(\mathfrak{n})$, we see that the closure of $\rho(Q')$ is compact. Using the exponential map $\exp : \mathfrak{n} \to N$ we see that there is a compact abelian group A of automorphisms of N and a homomorphism $Q' \to A$, also denoted ρ, such that $\rho(t')u = t' u t'^{-1}$ for every $t' \in Q'$ and $u \in N$. Define

$\rho^{-1} : Q' \to A$ by $\rho^{-1}(t') = [\rho(t')]^{-1}$. Let Γ the graph of ρ^{-1} . Now $Q \ltimes N$ has a compact presentation, hence so does $(Q \times A) \ltimes N$, by 1.1.2, hence so does $((Q \times A) \ltimes N)/\Gamma$ by 1.1.3 c), since Γ is a discrete closed central subgroup of $(Q \times A) \ltimes N$ isomorphic to $\mathbb{Z}^{n-1}$. The map $(t^n,u) \longrightarrow ((t^n,1),u)\Gamma$ embeds $<t> \ltimes N$ homeomorphically as a closed normal subgroup of $((Q \times A) \ltimes N)/\Gamma$ with factor group isomorphic to A . Therefore $<t> \ltimes N$ has a compact presentation, by 1.1.2.

We now may assume that Q is infinite cyclic, $Q = <t>$ say, $Q \ltimes N$ has a compact presentation and λ_+ and λ_- are elements of $R = \nu_* W(\mathfrak{n}^{ab})$ such that $\lambda_+(t) > 0$ and $\lambda_-(t) < 0$. We thus have a contradiction to 3.2.4 □

3.3 A COUNTEREXAMPLE

In the last section (theorem 3.2.3) we have seen that if $Q \ltimes N$ has a compact presentation, then there are no two opposite elements of R , i.e. if $\lambda \in S$ then $\{r\lambda ; r \leq 0\} \cap R = \emptyset$. This necessary condition for compact presentability is not sufficient if rank $Q = 2$. Even worse we give two groups N_1, N_2 with the same set R such that $Q \ltimes N_1$ has a compact presentation, whereas $Q \ltimes N_2$ does not.

3.3.1 EXAMPLE Let G_K be the group

$$G_K = \left\{ \begin{bmatrix} 1 & * & * & * \\ & * & * & * \\ & & * & * \\ & & & 1 \end{bmatrix} \right\}$$

i.e. $G_K \subset GL_4(K)$ is defined by the following equations, $a_{ij} = 0$ for $i > j$, $a_{11} = a_{44} = 1$. Let $N_K \subset G_K$ be the subgroup with $a_{ii} = 1$, $1 \leq i \leq 4$. Let $\pi \in K^*$ be an element with $\mathrm{mod}_K(\pi) < 1$. Let Q be the group of diagonal matrices of G_K with entries from $\langle\pi\rangle$. Let $\mathfrak{n}_K$ be the Lie algebra of upper triangular 4×4-matrices with entries from K and zeros on the diagonal. We have $N_K = \exp \mathfrak{n}_K$. The group Q acts on $\mathfrak{n}$ and N by conjugation. Define $d_i : Q \to K^*$, $i=1,\dots,4$, by mapping a matrix of Q to its i-th diagonal element. In the group $\mathrm{Hom}(Q,K^*)$, written additively, we have $d_1 = d_4 = 0$.

Let E_{ij} , $1 \leq i,j \leq 4$, be the 4×4-matrix with all entries zero except at the place where row i and column j intersect. E_{ij} is a basis of $\mathfrak{n}_K^{d_i - d_j}$, $1 \leq i < j \leq 4$. Set $\delta_i = \nu_*(d_i)$. So

$$R = \{-\delta_2, \delta_2 - \delta_3, \delta\} .$$

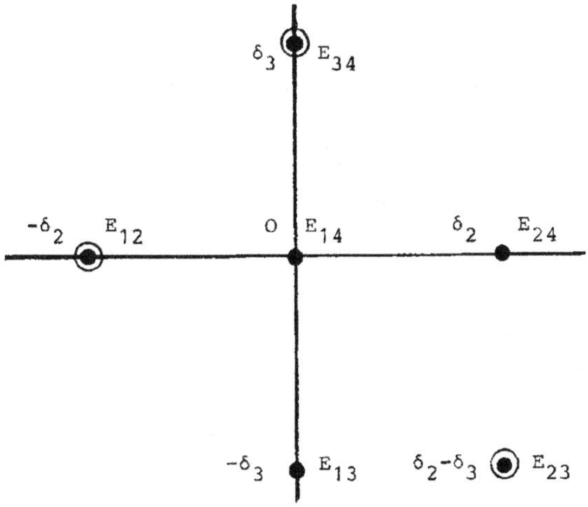

Let $\mathfrak{z}_K$ be the vector subspace of $\mathfrak{n}_K$ spanned by E_{14} . Then $\mathfrak{z}_K$ is a central ideal of $\mathfrak{n}_K$, centralized by Q . Define $\mathfrak{m}_K = \mathfrak{n}_K/\mathfrak{z}_K$. Then the natural homomorphism $\mathfrak{n}_K \to \mathfrak{m}_K$ induces an isomorphism $\mathfrak{n}_K^{ab} \xrightarrow{\sim} \mathfrak{m}_K^{ab}$, in particular

$$\nu_*(W(\mathfrak{n}_K^{ab})) = \nu_* W(\mathfrak{m}_K^{ab}) ,$$

i.e. the sets R for $\mathfrak{n}_K$ and $\mathfrak{m}_K$ coincide.

The point of the example is that $Q \ltimes \exp \mathfrak{n}_K$ has a compact presentation, whereas $Q \ltimes \exp \mathfrak{m}_K$ does not. The second claim follows from the fact that $Q \ltimes \exp \mathfrak{n}_K = Q \ltimes N$ has a compact set of generators, by 3.2.2, and $Z_K = \exp \mathfrak{z}_K$ is a closed central subgroup thereof which is not compactly generated, so $Q \ltimes \exp \mathfrak{m}_K = (Q \ltimes N_K)/Z_K$ is not compactly presentable, by 1.1.3 b). The first claim follows from the main theorem of this book (see 5.7.1). A different proof of the first claim for the case $K = \mathbb{Q}_p$ can be given as follows. For notations cf. 6.1. $G_{\mathbb{Z}[\frac{1}{p}]}$ has a finite presentation by [3]. So $G_{\mathbb{Q}_p}$ has a compact presentation by a result of Kneser [36]. Therefore $Q \ltimes N$ has a compact presentation by 1.1.2, since it is a closed normal subgroup of $G_{\mathbb{Q}_p}$ with compact factor group.

3.3.2 REMARK We shall see that the example $Q \ltimes M$ of a group that has no compact presentation but satisfies the necessary condition of 3.2.3, is typical in the following respect. There is a compactly presented locally compact topological group $Q \ltimes N$ and a surjection $Q \ltimes N \to Q \ltimes M$ whose kernel is not compactly generated as a normal subgroup of $Q \ltimes N$. We shall prove this in the next chapter. We shall see in chapter V that there is actually a Lie group N over K of this type (see 5.7.3). This will imply sufficiency of the conditions in our main result 5.6.1.

3.3.3 The group $\Gamma = G_{\mathbb{Z}[\frac{1}{p}]}$ is interesting in group theory. Its center, the upper right hand corner, is isomorphic to $\mathbb{Z}[\frac{1}{p}]$, hence not finitely generated. So this group gives a negative answer to the following question of P. Hall 1954 [31, p. 425]: Does every solvable finitely presented group satisfy the maximal condition for normal subgroups? Equivalently: Is every homomorphic image of a finitely presented solvable group itself finitely presented?

Another interesting fact about this group is that Γ modulo a cyclic subgroup of its center is not Hopfian and hence not residually finite.

For more consequences of this example for group theory see [3] and [49 , sections 3.4, 4.2 and 8].

3.4 The geometric invariant of Bieri and Strebel

In this section we recall the definition of the geometric invariant of Bieri and Strebel, give an obvious generalization for topological groups and show how it is related to the set R of 3.1.6.

Let Q be a finitely generated abelian group acting by automorphism on the abelian group A. For every $\lambda \in \mathrm{Hom}(Q,\mathbb{R})$ define the monoid $Q_\lambda = \{q \in Q ; \lambda(q) \geq 0\}$. Bieri and Strebel have defined the following geometric invariant [13] of the $\mathbb{Z}[Q]$-module A.

$$\Sigma_A = \{\lambda \in \mathrm{Hom}(Q,\mathbb{R}) ; \lambda \neq 0 \text{ and } A \text{ is finitely generated as a module over the monoid algebra } \mathbb{Z}[Q_\lambda]\}/\sim ,$$

where $\sim$ is the equivalence relation whose equivalence classes are the open rays in $\mathrm{Hom}(Q,\mathbb{R})$ starting at 0. So Σ_A may be considered as a subset of the sphere $S(Q)$ in $\mathrm{Hom}(Q,\mathbb{R})$.

An abelian topological group A together with an action of Q on A by topological automorphisms will be called a *topological Q-module.* One has an obvious generalization of the invariant Σ_A for locally compact topological Q-modules obtained by replacing "finitely generated" by "compactly generated" in the definition of Σ_A (compare 7.3.1).

3.4.1 Lemma *If* $A \xrightarrow{\iota} B \xrightarrow{\pi} C$ *is an exact sequence of locally compact topological Q-modules then*

$$\Sigma_B = \Sigma_A \cap \Sigma_C \quad \square$$

The hypothesis of exactness is shorthand for the following conditions: $\iota : A \to B$ is a topological isomorphism of A onto a closed subgroup of B and $\pi : B \to C$ is an open surjective homomorphism with kernel $\iota(A)$.

It is sometimes more convenient to deal with the complement

$$\Sigma_A^c = S(Q) \setminus \Sigma_A \quad .$$

3.4.2 LEMMA *Let* K *be a p-adic field, let* $\nu = -\log \circ \mathrm{mod}_K$ *be the natural exponential valuation of* K . *Suppose an action of* Q *on* K *is given by a homomorphism* $u : Q \to K^*$. *Then*

$$\Sigma_K^c = \begin{cases} \emptyset & \text{if } \nu_*(\mu) = 0 \\ \nu_*(\mu)/\sim & \text{if } \nu_*(\mu) \neq 0 \end{cases} .$$

PROOF. Suppose $0 \neq \lambda \in \mathrm{Hom}(Q,\mathbb{R})$. If $\nu_*(\mu)|Q_\lambda \geq 0$ and C is a compact subset of K then $\{x \in K ; \nu(x) \geq \inf \nu(C)\}$ is a compact $\mathbb{Z}[Q_\lambda]$ submodule of K . So $\Sigma_K^c = \emptyset$ if $\nu_*(\mu) = 0$, and $\nu_*(\mu)/\sim \in \Sigma_K^c$ if $\nu_*(\mu) \neq 0$. Now let $0 \neq \lambda \in \mathrm{Hom}(Q,\mathbb{R})$ not be a positive multiple of $\nu_*(\mu) \neq 0$. We have to show that $\lambda \in \Sigma_K$. Let $V = \mathrm{Hom}(Q,\mathbb{R})$. Evaluating at the elements of Q defines a homomorphism $Q \to V^*$ whose image is a lattive Γ in the dual space V^* of V . Since λ is not a positive multiple of $\nu_*(\mu)$, there is an element $\ell \in V^*$ such that $\ell(\nu_*(\mu)) < 0$ and $\ell(\lambda) > 0$, by corollary A.2 1) $\Rightarrow$ 3) of the appendix applied to $T = \{\lambda, -\nu_*(\mu)\}$. So in the dense subspace $\Gamma \otimes \mathbb{Q}$ of V^* there is an element ℓ satisfying these inequalities, hence there is one in Γ . So finally there is an element $t \in Q$ such that $\nu_*(\mu)(t) < 0$ and $\lambda(t) > 0$. This element $t \in Q_\lambda$ acts expanding on K . So K is the only $\mathbb{Z}[Q_\lambda]$-submodule of K containing some neighborhood of 0 , hence $\lambda \in \Sigma_K$ □

So Bieri and Strebel's Σ_{nab}^c is the same as our R modulo $\sim$, as follows from 3.4.1 and 3.4.2:

3.4.3 COROLLARY *With the hypotheses and notations of 3.2 suppose* $0 \notin R$. *Then* $\Sigma_{\text{nab}}^c = R/\sim$ □

So theorem 3.2.3 is analogous to a result of [14] giving a necessary condition for finite presentability of arbitrary solvable groups.

IV. IMPLICATIONS OF THE NECESSARY CONDITION

In this chapter we suppose the necessary condition for compact presentability satisfied (namely condition 1) of 3.2.3, equivalently tameness of the representation of Q on $\mathfrak{n}^{ab}$. We define a certain topological group, H_U of 4.5. It has a compact presentation, by definition. The main result of this chapter is, that the natural map $\pi_U : H_U \to Q \ltimes N$ is close to a central extension. More precisely, the kernel of π_U is central in $\pi_U^{-1}(N)$ and every one of its elements has a finite conjugacy class (see 4.6). This result will be used in the next chapter to show that condition 1) together with a certain condition 2) are sufficient for π_U to be an isomorphism. Condition 2) will be stated in terms of central extensions not of the group N but of its Lie algebra $\mathfrak{n}$. Technically speaking it is a condition about the second homology of the Lie algebra $\mathfrak{n}$. This condition about the Lie algebra homology should be easy to check.

Another pleasant feature of π_U is that either π_U is an isomorphic - and hence $Q \ltimes N$ has a compact presentation - or $Q \ltimes N$ has no compact presentation. So either our construction H_U gives a compact presentation of $Q \ltimes N$ or there is none at all.

Here is a survey of the contents of the different sections. We continue to use the notations and hypotheses of 3.1. In 4.1 we define our building blocks for presenting $Q \ltimes N$, namely certain subgroups $Q \ltimes N^c$ of $Q \ltimes N$ such that N^c admits a contracting element $t \in Q$. In 4.2 we recall the notion of the colimit of a functor with values in the category of groups. We define $H = Q \ltimes M$, where M is the colimit of the N^C's. The most difficult part of the proof will be to prove that M is nilpotent, if condition 1) holds. This is proved in 4.4. To do this we need formulae on commutators in $\mathfrak{n}$, which are proved in 4.3. Even if $Q \ltimes N$ has a compact presentation, the natural map $Q \ltimes M \to Q \ltimes N$ is not an isomorphism, in general. An example will be given later on (see 5.7.4). One has to take into account the topology. So in 4.5 further relations are introduced to turn H into a locally compact topological group H_U locally isomorphic to $Q \ltimes N$. The group H_U has a compact presentation, actually its definition may be rewritten as an explicit compact presentation. Finally in 4.6 we prove the dichotomy mentioned above: Either $H_U \xrightarrow{\sim} Q \ltimes N$ or $Q \ltimes N$ has no compact presentation. We also give the information about the kernel of $H_U \to Q \ltimes N$ mentioned above. In the next chapter we shall translate the existence of such an extension $H_U \to Q \ltimes N$ into a handier condition, formulated in terms of the second homology of $\mathfrak{n}$. In 4.7 we study the Lie algebra analogue $\varinjlim \mathfrak{n}^C$ of M, which is much easier. This Lie algebra analogue can serve as a motivation and guideline for the group theoretical computations. The reader may want to look into the Lie algebra case before studying the group case, as the author did.

4.1 The building blocks N^C

In this section we define the building blocks our our theory, the groups $Q \ltimes N^C$.

The hypotheses and notations of 3.1 will be assumed, in particular $\mathfrak{n} = \oplus\, \mathfrak{n}^\lambda$, $\lambda \in W(\mathfrak{n})$.

Let C be a subsemigroup of $\mathrm{Hom}(Q,\mathbb{R})$. Then

(4.1.1) $$\mathfrak{n}^C := \oplus\, \mathfrak{n}^\lambda, \quad \nu_*(\lambda) \in C$$

is a Lie subalgebra of $\mathfrak{n}$, by 3.1.1. Define

$$N^C = \exp \mathfrak{n}^C$$

a closed Lie subgroup of N normalized by Q.

Let $t \in Q$. Define the subsemigroup

(4.1.2) $$C_t = \{\mu \in \mathrm{Hom}(Q,\mathbb{R}) \; ; \; \mu(t) > 0\} .$$

These semigroups are important for us because $t \in Q$ is contracting on N^{C_t}. Here is a simple necessary and sufficient condition to see when N^C is contained in N^{C_t} for some $t \in Q$ and hence has a compact presentation.

4.1.3 Lemma *Let C be a subsemigroup of $\mathrm{Hom}(Q,\mathbb{R})$. Suppose $0 \notin C$. Then there is an element $t \in Q$ such that $C \cap \nu_* W(\mathfrak{n}) \subset C_t$. It follows that t is contracting on N^C and hence that $Q \ltimes N^C$ has a compact presentation.*

Proof. Recall from 3.1.4 that $\nu_* : \mathrm{Hom}(Q,K^*) \to \mathrm{Hom}(Q,\mathbb{R})$, where $\nu_*(\lambda) = -\log \circ \mathrm{mod}_K \circ \lambda$. By choosing an appropriate basis for the logarithm we may assume that the image of ν_* is $\mathrm{Hom}(Q,\mathbb{Z})$. Then $V = \mathrm{Hom}(Q,\mathbb{Q})$ is the $\mathbb{Q}$-vector space in $\mathrm{Hom}(Q,\mathbb{R})$ spanned by the image of ν_*. If now $0 \notin C$, there is a linear form $\ell \in V^*$ such that $\ell|\nu_*(W(\mathfrak{n})) \cap C > 0$ by corollary A.2,1) $\Rightarrow$ 2) of the appendix, applied to $T = \nu_*(W(\mathfrak{n})) \cap C$. The image of the evaluation map $Q \longrightarrow V^*$ contains a basis of the $\mathbb{Q}$-vector space V^*, hence some non zero integer multiple of ℓ is in the image of Q, in other words, there is an element $t \in Q$ such that $\nu_*(\lambda)(t) > 0$ for every $\nu_*(\lambda) \in \nu_*(W(\mathfrak{n})) \cap C$, which implies our first claim.

Now $\langle t\rangle \ltimes N^C$ has a compact presentation by 1.3.1, hence so does $Q \ltimes N^C$ by 1.1.3a), since $(Q \ltimes N^C)/(\langle t\rangle \ltimes N^C) = Q/\langle t\rangle$ is a finitely generated abelian group and hence has a finite presentation, which can be shown e.g. by induction on the number of generators using 1.1.3a) □

4.2 THE COLIMITS H AND M

In this section we define our basic colimits $H = \varinjlim Q \ltimes N^C$ and $M = \varinjlim N^C$ (see 4.2.8) and give some elementary properties thereof. We first recall the notion of the colimit of a functor and state a sufficient condition for compact presentability of a colimit of topological group (4.2.6).

4.2.1 Let I and C be categories and let $f : I \to C$ be a functor. A *colimit* of f, denoted $\varinjlim f$, is an object C of C together with a natural transformation F from the functor f to the constant functor $I \to C$ with value $(C, 1_C)$ having the following universal property. Given an object C' of C and a natural transformation F' from f to the constant functor $I \to C$ with value $(C', 1_{C'})$, there is exactly one arrow $h : C \to C'$ in C such that $h \circ F(i) = F'(i)$ for every object i of I. If a colimit of f exists, it is unique up to unique isomorphism.

4.2.2 Given a functor $\varphi : J \to I$ and a functor $f : I \to C$ and colimits $\varinjlim f = (C, F)$ and $\varinjlim f \circ \varphi = (C', F')$, then there is a unique arrow $h : C' \to C$, denoted $\varinjlim \varphi : \varinjlim f \circ \varphi \to \varinjlim f$ such that $h \circ F'(j) = F(\varphi(j))$ for every object j of J. We look for conditions that $\varinjlim \varphi$ is an isomorphism.

4.2.3 Let $\varphi : J \to I$ be a functor. If i is a fixed object of I, following Quillen [42], we denote by $i\backslash\varphi$ *the category of objects over* i whose objects are pairs (j, v), where J is an object of J and $v : i \to \varphi(j)$ is an arrow in I and whose morphisms from (j, v) to (j', v') are arrows $w : j \to j'$ in J such that $\varphi(w) \circ v = v'$.

Recall that a category is *connected* if for every two objects C and C' there is a sequence of objects and arrows as follows

$C = C_0 \to C_1 \leftarrow C_2 \to C_3 \leftarrow C_4 \to \dots \to C_n = C'$. With these notations we have the following lemma.

4.2.4 LEMMA $\varinjlim \varphi : \varinjlim f\circ\varphi \to \varinjlim f$ *is an isomorphism if* $i\backslash\varphi$ *is non-empty and connected for every object* i *of* I .

PROOF. Let $\varinjlim f = (C,f)$ and $\varinjlim f\circ\varphi = (C',F')$. We define an arrow $g : C \to C'$ by giving a natural transformation G from $f : I \to C$ to the constant functor with value C' . Let i be an object of I . There is an object j of J and an arrow $v : i \to \varphi(j)$, since $i\backslash\varphi$ is not empty. Define $G(j,v) : f(i) \to C'$ to be the composition of $f(v) : f(i) \to f(\varphi(j))$ and $f'(j) : f\circ\varphi(j) \to C'$. For every morphism $(j,v) \to (j',v')$ in $i\backslash\varphi$ we obtain a commutative diagram showing that $G(j,v) = G(j',v')$. Hence if $i\backslash\varphi$ is non-empty and connected we thus obtain a well defined arrow $G(i) : f(i) \to C'$. It defines a natural transformation G as claimed, since every arrow $i \to i'$ induces a functor $i'\backslash\varphi \to i\backslash\varphi$. The two arrows, $\varinjlim \varphi$ and the arrow g induced by G , are easily seen to be inverse of each other □

4.2.5 Recall the construction of a colimit of a functor f from a small category I to the category $\mathcal{G}r$ of groups. Take the free product $\coprod f(i)$ of the groups $f(i)$, $i \in I$, and divide out the smallest normal subgroup containing all the elements of the form $x^{-1}\cdot f(h)(x)$, where $h : i \to i'$ is an arrow in I and $x \in f(i)$.

Suppose we have a presentation $\langle X_i, R_i\rangle$ for every group $f(i)$, $i \in I$. Then $\left\langle \coprod_i X_i ; \bigcup_{i\in I} R_i \cup \bigcup_{h\in \mathrm{Hom}(I)} P_h \right\rangle$ is a presentation of $\varinjlim f$, where for $h : i \to i'$ in $\mathrm{Hom}(I)$ the set P_h is a set of elements $x\cdot y^{-1} \in F(\coprod_i X_i)$ obtained by choosing for every element $x \in X_i$ an element $y \in F(X_{i'})$ such that the image of y under $F(X_{i'}) \to f(i')$ is $f(h)(x)$.

4.2.6 LEMMA *Let* I *be a finite actegory, i.e.* mor I *(and hence* ob I*) is a finite set. Let* f *be a functor from* I *to the category of locally compact topological groups. Suppose there is a colimit* $\varinjlim f$ *in the category of locally compact topological groups, such that, after forgetting topologies, it is also a colimit in the category of groups. Then if* f(i) *has a compact presentation for every* $i \in I$ *, so does* $\varinjlim f$.

PROOF. Let $\langle X_i, R_i \rangle$ be a presentation of f(i) with X_i compact and R_i of bounded reduced length (see 1.1.1). Then the above presentation of $\varinjlim f$ proves our claim, when we have shown that we can choose every P_h of finite reduced length. This is implied by the following lemma.

4.2.7 LEMMA *Let* G *be a locally compact topological group and let* X *be a compact set of generators of* G . *Then for every compact subset* Y *of* G *there is a number* $n \in \mathbb{N}$ *such that* $Y \subset (X \cup X^{-1})^n = \{x_1^{\varepsilon_1} \ldots x_m^{\varepsilon_m} ; m \leq n , x_i \in X , \varepsilon_i \in \{+1,-1\}\}$.

PROOF. Define $X_n = (X \cup X^{-1})^n$ for $n \in \mathbb{N}$. By hypothesis $\cup X_n = G$. So by the Baire category theorem some X_m contains a non-empty open set, hence $X_{2m} = X_m \cdot X_m^{-1}$ contains a neighborhood of e , hence X_{n+2m} is a neighborhood of X_n , hence $\cup \mathring{X}_n = G$. A finite number of $\mathring{X}_n$ covers Y , hence $Y \subset \mathring{X}_n$ for some $n \in \mathbb{N}$ □

I will turn out that, for the groups we are interested in, we can either prove compact presentability by this simple observation 4.2.6 or our group has no compact presentation at all, see 4.6.2

4.2.8 The group we shall study is the following one. We make the assumptions of 4.1. Let $\mathcal{C}$ be the set of all subsemigroups C of Hom(Q,$\mathbb{R}$) *not containing zero*, partially ordered by inclusion. Regard $\mathcal{C}$ as a category in the usual way. So the objects of $\mathcal{C}$ are the elements of $\mathcal{C}$. The set of arrows in $\mathcal{C}$ from C_1 to C_2 is empty if

$C_1 \not\subset C_2$ and contains exactly one element - denoted $C_1 \subset C_2$ - if $C_1 \subset C_2$. Define functors $\mathcal{C} \to \mathcal{G}r$ by $C \longmapsto Q \ltimes N^C$ and $C \longmapsto N^C$, inclusion going to the embedding homomorphism. Let H and M be the resp. colimits. We write

$$M = \varinjlim_{\mathcal{C}} N^C \ , \ H = \varinjlim_{\mathcal{C}} (Q \ltimes N^C) = Q \ltimes M \ .$$

The embeddings $Q \ltimes N^C \to Q \ltimes N$ induce a group homomorphism

$$\pi : Q \ltimes M \to Q \ltimes N \ .$$

The maps $\varphi^C : Q \ltimes N^C \to Q \ltimes M = H$, which are part of the definition of the colimit, compose to give a map $\varphi : \bigcup Q \ltimes N^C \to H$, such that

$$\varphi | Q \ltimes N^C = \varphi^C : Q \ltimes N^C \to H$$

is a group homomorphism for every $C \in \mathcal{C}$. The group H, and similarly M, can also be described as the universal group admitting such a map φ.

Instead of $\mathcal{C}$ we may take the finite ordered set $\mathcal{C}_1 = \{C \cap \nu_* W(\mathfrak{n})\ ;\ C \in \mathcal{C}\}$ and obtain the same colimits, obviously. Hence if $\pi : M \to N$ is an isomorphism, $Q \ltimes N$ has a compact presentation by 4.2.6 and 4.1.3.

Define $\mathcal{D}_1 = \{C_t \cap C_{t'} \cap \nu_*(W(\mathfrak{n}))\ ;\ t,t' \text{ in } Q\}$. The inclusion $\mathcal{D}_1 \subset \mathcal{C}_1$ satisfies the hypothesis of 4.2.4 by 4.1.3 (note that $t = t'$ is not forbidden), hence $H = \varinjlim_{\mathcal{D}} N_C$, and $H = \varinjlim_{\mathcal{D}_1} (Q \ltimes N^C)$. So M and H are the free product of the groups N^{C_t} and $Q \ltimes N^{C_t}$ resp., amalgamated along their intersections.

$$M = \coprod_{\cap} N^{C_t} \ , \ H = \coprod_{\cap} (Q \ltimes N^{C_t}) = Q \ltimes M \ , \ t \in Q \ .$$

4.2.9 LEMMA $\pi : Q \ltimes M \to Q \ltimes N$ *is surjective iff* $0 \notin R = \nu_* W(\mathfrak{n}^{ab})$.

Recall that this is equivalent to compact generation of $Q \ltimes N$ by 3.2.2.

PROOF. $\pi : Q \ltimes M \to Q \ltimes N$ is surjective iff $\bigcup N^C = \bigcup N^{C_t}$ generates N iff $\bigcup \mathfrak{n}^{C_t}$ generates $\mathfrak{n}$, by 2.5.9, iff the union of the images of $\mathfrak{n}^{C_t}$ in $\mathfrak{n}^{ab}$ generates $\mathfrak{n}^{ab}$, by 2.1.3, iff $0 \notin R$, obviously □

4.3 Commutators in $\mathfrak{n}$

The main result of this chapter will be that M is nilpotent, if any two elements of R are positively independent. This will be proved in 4.4. For that proof we need commutator relations in $\mathfrak{n}$. Such commutator relations are proved in this section.

What we are really interested in are commutator relations in M. We have $(\varphi(x),\varphi(y)) = \varphi(x,y)$ if $x \in \exp \mathfrak{n}^{\alpha}$, $y \in \exp \mathfrak{n}^{\beta}$ and α and β are not opposite of each other, i.e. there is no positive linear dependence between α and β, because then $0 \notin C = \{r_1\alpha + r_2\beta,\ r_1 \geq 0,\ r_2 \geq 0,\ r_1 + r_2 > 0\}$ and we have a homomorphism $\varphi^C = \varphi | N^C : N^C \to M$. So in order to get hold of commutator relations in M, we shall make use of a result about commutators in $\mathfrak{n}$ of the following type. "*Enough commutator relations in* $\mathfrak{n}$ *follow from commutator relations between* $\mathfrak{n}^{\alpha}$*'s and* $\mathfrak{n}^{\beta}$*'s as above.*" An information of this type is contained in lemma 4.3.1. It will be applied in 4.4 and 4.7.

If A is a subset of a real vector space, let us denote by $\operatorname{conv}(A)$ the convex hull of A, i.e. the set of all positive linear combinations $\Sigma \lambda_a \cdot a$ with $\lambda_a \geq 0$ and $\Sigma \lambda_a = 1$. If $A = \{x_1,\dots,x_n\}$ we write $\operatorname{conv}(x_1,\dots,x_n)$ instead of $\operatorname{conv}\{x_1,\dots,x_n\}$. Note that the set A is positively independent iff $0 \notin \operatorname{conv}(A)$.

Let $\mathfrak{n} = \mathfrak{n}_1 \supset \mathfrak{n}_2 \supset \dots$ be the descending central series of $\mathfrak{n}$. For $\alpha \in \operatorname{Hom}(Q,\mathbb{R})$ define

$$\mathfrak{n}_i^{\alpha} = \mathfrak{n}^{\alpha} \cap \mathfrak{n}_i \quad .$$

4.3.1 Lemma *Suppose any two elements of* $R = \nu_* W(\mathfrak{n}^{ab})$ *are positively independent. Then we have for every integer* $i \geq 2$ *and every*

$\alpha \in \mathrm{Hom}(Q,\mathbb{R})$ *the following equations.*

a) For $\alpha \neq 0$

$$\mathfrak{n}_i^{\alpha} = \Sigma\,[\mathfrak{n}_k^{\beta}, \mathfrak{n}_{\ell}^{\gamma}]$$

where the sum is extended over all β,γ,k *and* ℓ *such that* $\alpha = \beta + \gamma$, $0 \notin \mathrm{conv}(\beta,\gamma)$, $k+\ell = i$.

b) For $\alpha \neq 0$ *and* $\mathbb{R}_+^*\alpha \cap R = \emptyset$

$$\mathfrak{n}_i^{\alpha} = \Sigma\,[\mathfrak{n}_k^{\beta}, \mathfrak{n}_{\ell}^{\gamma}]$$

with $\beta + \gamma = \alpha$, β,γ *not in* $\mathbb{R}\alpha$, $k + \ell = i$.

c) If $\alpha_o \neq 0$, *we have for* $0 \in \mathrm{Hom}(Q,\mathbb{R})$

$$\mathfrak{n}_i^{o} = \Sigma\,[\mathfrak{n}_k^{\beta}, \mathfrak{n}_{\ell}^{-\beta}]$$

with $\beta \notin \mathbb{R}\alpha_o$, $k + \ell = i$.

Here we denote by $[V,W]$ for two vector subspaces V,W of a Lie algebra $\mathfrak{g}$ the vector space (= $\mathbb{Z}$-module) spanned by the set of brackets $[v,w]$, $v \in V$, $w \in W$.

Parts a) and b) of the lemma will be used at two decisive points of the next section, c) is only used for the proof of 4.3.1.

I want to give a short explanation of what the content of the lemma is. For $i = 2$ part a) says that a commutator in $\mathfrak{n}^{\alpha}$, $\alpha \neq 0$, is a sum of commutators, each within one $\mathfrak{n}^{C}$, $C \in \mathcal{C}$, and analogously for high-

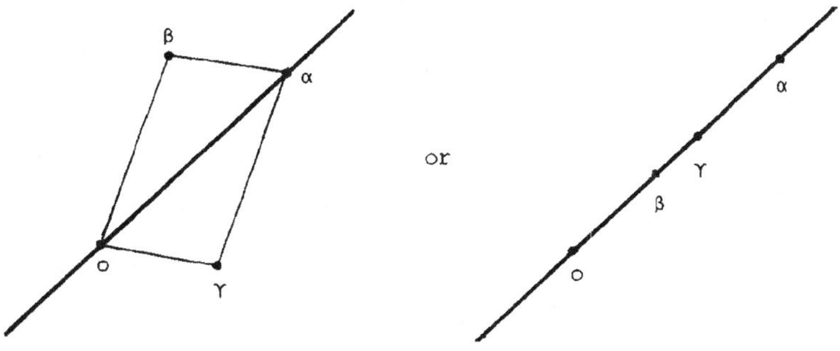

er commutators. In part b) the additional hypothesis that the ray through α does not intersect R allows one to exclude the case given by the right hand figure. Finally in part c) it is shown that for $\alpha = 0$ one can exclude β's and γ's on a given line.

PROOF. We shall make frequent use of the following trivial remark. Any two elements of R are positively independent iff the following two conditions hold i) $0 \notin R$, ii) there are no two opposite elements of R , i.e. if the ray $\mathbb{R}_+^*\alpha = \{r\alpha ; r > 0\}$ intersects R for $\alpha \neq 0$, then the oppositve ray $-\mathbb{R}_+^*\alpha$ does not intersect R .

We prove the lemma by induction on the step s of nilpotence of $\mathfrak{n}$. For $s = 1$ there is nothing to prove. Suppose the lemma is true for all nilpotent Lie algebras of step $< s$ satisfying the assumptions of 3.1. Define $\mathfrak{m} = \mathfrak{n}/\mathfrak{n}_s$. It suffices to prove the lemma for $i = s$, because the exact sequence

$$0 \to \mathfrak{n}_s^\alpha \to \mathfrak{n}_i^\alpha \to \mathfrak{m}_i^\alpha \to 0$$

together with the inductive hypothesis and the trivial fact that the right hand sinde of every one of the claimed equations is contained in the left hand side, implies the result for every $i \geq 2$ if it is proved for $i = s$.

If X is a set of generators of $\mathfrak{n}$ and $\mathfrak{a}$ is an ideal of $\mathfrak{n}$, then $[\mathfrak{n},\mathfrak{a}] = \sum_{X \in X} ad(X)\mathfrak{a}$, because the right hand side is a vector space V contained in $[\mathfrak{n},\mathfrak{a}]$ and the set of elements $Y \in \mathfrak{n}$ such that $ad(Y)\mathfrak{a} \subset V$ forms a Lie algebra, since $V \subset \mathfrak{a}$, and contains a set of generators of $\mathfrak{n}$, hence $[\mathfrak{n},\mathfrak{a}] \subset V$.

The Lie algebra $\mathfrak{n}$ is generated by $\cup\, \mathfrak{n}^\beta$, $\beta \in R$. It follows that $\mathfrak{n}_s^\alpha = \Sigma[\mathfrak{n}^\beta,\mathfrak{n}_{s-1}^\gamma]$, $\beta + \gamma = \alpha$, $\beta \in R$. So we have to prove that

$$[\mathfrak{n}^\beta,\mathfrak{n}_{s-1}^\gamma] , \beta + \gamma = \alpha , \beta \in R$$

is contained in the resp. right hand side of 4.3.1. This is trivially true if β and γ are linearly independent. If β and γ are linearly dependent, we have $\gamma = r\beta$, since $\beta \in R$ is $\neq 0$. There are three cases to consider.

<u>$r > 0$</u> $\alpha = \beta + \gamma \neq 0$. We are not in case b) nor c) of 4.3.1, and $[n^{\beta}, n^{\gamma}_{s-1}]$ is a summand in a), since $0 \notin \mathrm{conv}(\beta,\gamma)$.

<u>$r = 0$</u>, i.e. $\gamma = 0$, $\alpha = \beta \neq 0$. The inductive hypothesis 4.3.1 c) implies with $\alpha_o = \beta$ that

$$n^{o}_{s-1} \subset \Sigma\, [n^{\delta}_{k}, n^{-\delta}_{\ell}] + n_s \ , \ \delta \notin \mathbb{R}\beta \ , \ k + \ell = s - 1 \ .$$

Since n_s is central in n , the Jacobi identity implies that

$$[n^{\beta}, n^{o}_{s-1}] \subset \Sigma\, [n^{\beta+\delta}_{k+1} , n^{-\delta}_{\ell}] + \Sigma\, [n^{\delta}_{k}, n^{\beta-\delta}_{\ell}]$$

with $\beta + \delta$, δ , $\beta - \delta$ not in $\mathbb{R}\beta$. We thus have the assertion of b), which is stronger than that of a).

<u>$r < 0$</u> $\alpha = \beta + \gamma$, $\mathbb{R}^*_+\gamma \cap R = \mathbb{R}^*_+(-\beta) \cap R = \emptyset$, by the first remark of this proof, since $\beta \in R$. The inductive hypothesis implies that

$$n^{\gamma}_{s-1} \subset \Sigma\, [n^{\delta}_{k} , n^{\varepsilon}_{\ell}] + n^{\gamma}_{s} , \ \ \delta,\varepsilon \notin \mathbb{R}\beta = \mathbb{R}\gamma \ , \ \delta+\varepsilon = \gamma \ , \ k + \ell = s - 1 \ .$$

Hence

$$[n^{\beta}, n^{\gamma}_{s-1}] \subset \Sigma\, [n^{\beta+\delta}_{k+1} , n^{\varepsilon}_{\ell}] + \Sigma\, [n^{\delta}_{k} , n^{\beta+\varepsilon}_{\ell+1}]$$

with $\beta + \delta$, ε , δ , $\beta + \varepsilon$ not in $\mathbb{R}\beta = \mathbb{R}\gamma$. This proves the assertion b), hence a), if $\alpha \neq 0$ and c) if $\alpha = 0$ and $\alpha_o \in \mathbb{R}\beta$. The case $\alpha = \beta + \gamma = 0$ and $\alpha_o \notin \mathbb{R}\beta = \mathbb{R}\gamma$ is trivial, anyway □

4.4 THE DESCENDING CENTRAL SERIES OF M

In this section we prove the main result of this chapter, namely that M is nilpotent. The proof runs roughly as follows. Let M_i be the descending central series of M . The first step (4.4.12) is that M_i is generated by the special commutators concentrated on lines, the M^L_i of 4.4.10. The second step 4.4.13 is that a given line L_o can be excluded from the set of complicated generators M^L_i . This then implies the result very easily (4.4.14). An important part of the proof is to translate the result of 4.3 about commutators in the Lie algebra $\mathfrak{n}$ into a result (4.4.7) about commutators in the group M. That lemma is used for excluding a given line from the set of generators of M_i. We often use the formula of P. Hall's (2.2.3,3), a group theoretical analogue of the Jacobi identity. A difficulty arises from the conjugations in P. Hall's formula. So we often have to deal with normal subgroups instead of single elements. This makes the translation of Lie algebra results into group results sometimes tedious and awkward.

Recall from 4.2.8 the following definitions and notations. Let $\mathcal{C}$ be the ordered set of subsemigroups C of $\mathrm{Hom}(Q,\mathbb{R})$ not containing zero. Let $M = \varinjlim N^C$, $H = Q \ltimes M = \varinjlim (Q \ltimes N^C)$ the limit being taken in the category of groups and over the ordered set $\mathcal{C}$ regarded as a category. We have a map $\varphi : \cup N^C \to M$ such that $\varphi | N^C : N^C \to M$ is a group homomorphism for every $C \in \mathcal{C}$. The inclusions $N^C \to N$ induce a group homomorphism $\pi : M \to N$ such that $\pi \circ \varphi = \mathrm{Id}$ on $\cup N^C$.

The aim of this section is to show that M is nilpotent, if there are not two opposite elements of R. The much easier Lie algebra analogue is proved in 4.7.

Let $\mathfrak{n} = \mathfrak{n}_1 \supset \mathfrak{n}_2 \supset \dots$ be the descending central series of $\mathfrak{n}$. Recall the notation

<u>(4.4.1)</u> $$\mathfrak{n}_i^\lambda = \mathfrak{n}^\lambda \cap \mathfrak{n}_i$$

for $\lambda \in \mathrm{Hom}(Q,\mathbb{R})$. For $C \in \mathcal{C}$ set

<u>(4.4.2)</u> $$\mathfrak{n}_i^C := \oplus\, \mathfrak{n}_i^\lambda\,,\ \lambda \in C\,.$$

$\mathfrak{n}_i^C$ is an ideal of $\mathfrak{n}^C = \mathfrak{n}_1^C$, therefore

<u>(4.4.3)</u> $$N_i^C := \exp \mathfrak{n}_i^C$$

is a normal subgroup of N^C and

<u>(4.4.4)</u> $$M_i^C := \varphi(N_i^C)$$

is a normal subgroup of $M^C := M_1^C$.

Let us warm up with an easy lemma.

4.4.5 LEMMA *If* C_1, C_2 *and* $C_1 + C_2 = \{c_1 + c_2 ; c_i \in C_i\}$ *are in* $\mathcal{C}$ *, then*

$$(M_j^{C_1}, M_k^{C_2}) \subset M_{j+k}^{C_1+C_2} .$$

PROOF. $C = C_1 \cup C_2 \cup C_1 + C_2$ is a semigroup not containing 0 . Now $\mathfrak{n}_j^{C_1 \cup (C_1+C_2)} := \mathfrak{a}$ and $\mathfrak{n}_k^{C_2 \cup (C_1+C_2)} =: \mathfrak{b}$ are ideals of $\mathfrak{n}^C$, hence $(\exp \mathfrak{a}, \exp \mathfrak{b}) \subset \exp \mathfrak{n}_{j+k}^{C_1+C_2}$ by 2.5.12, which implies our claim by applying φ^C □

An open ray in $\mathrm{Hom}(Q,\mathbb{R})$ is a subset of $\mathrm{Hom}(Q,\mathbb{R})$ of the form $\{rv ; r > 0\}$ for some $v \neq 0$. Let $\mathcal{S}$ be the set of open rays in $\mathrm{Hom}(Q,\mathbb{R})$. The letter $\mathcal{S}$ is chosen because there is a bijection between $\mathcal{S}$ and the sphere in $\mathrm{Hom}(Q,\mathbb{R})$. We have $\mathcal{S} \subset \mathcal{C}$ and

(4.4.6) $$M_i^C = \langle \cup M_i^{C \cap S} , S \in \mathcal{S} \rangle$$

since the same holds for N_i^C by 2.5.9.

Here are the two consequences of the lemma in 4.3 we shall need.

4.4.7 LEMMA *Suppose any two elements of* R *are positively independent. Let* S *be a ray in* $\mathrm{Hom}(Q,\mathbb{R})$ *and* $i \geq 2$.

a) $M_i^S \subset \langle \cup (M_j^C, M_k^C) ; j + k = i , S \subset C \in \mathcal{C} \rangle$.

b) If $S \cap R = \emptyset$ *, then* $M^S = M_2^S$ *and*

$M_i^S \subset \langle \cup (M_{j_1}^{S_1}, M_{j_2}^{S_2}) ; j_1 + j_2 = i , S_1, S_2 \text{ in } \mathcal{S} , S \subset S_1 + S_2, S_1 \not\subset \text{span } S , S_2 \not\subset \text{span } S \rangle$.

PROOF. a) $(N_j^C, N_k^C) = \exp[\mathfrak{n}_j^C, \mathfrak{n}_k^C]$ by 2.5.12 applied to the ideals $\mathfrak{n}_j^C$ and $\mathfrak{n}_k^C$ of $\mathfrak{n}^C$. In particular we have $\exp \mathfrak{n}_{j,k,C} \subset (N_j^C, N_k^C)$ for $\mathfrak{n}_{j,k,C} := \mathfrak{n}_i^S \cap [\mathfrak{n}_j^C, \mathfrak{n}_k^C]$, if $i = j+k$ and $S \subset C \in \mathcal{C}$. If we let j,k and C run through all such combinations the family of $\mathfrak{n}_{j,k,C}$ generates $\mathfrak{n}_i^S$ as a $\mathbb{Z}$-module by 4.3.1 a). So the set of elements $x \in N_i^S$

such that $\varphi^S(x) \in A := \langle \cup (M_j^C, M_k^C) ; j+k=i , S \subset C \in \mathcal{C} \rangle$ is a group containing all the $\exp \mathfrak{n}_{j,k,C}$, hence equals N_i^S by 2.5.9, so $M_i^S \subset A$.

b) The first claim follows immediately from the definition of $R = \nu_* W(\mathfrak{n}^{ab})$.

Set $L = \text{span } S$, the line spanned by S . Let S_1 and S_2 be open rays in $\text{Hom}(Q,\mathbb{R})$, $S \subset S_1 + S_2$, $S_1 \not\subset L$, $S_2 \not\subset L$. Then $C = S_1 \cup S_2 \cup S_1 + S_2 \in \mathcal{C}$. Set $\mathfrak{a}_\ell = \mathfrak{n}_{j_\ell}^{S_\ell}$, $\ell = 1,2$. Let $\mathfrak{h} = \langle \mathfrak{a}_1 \cup \mathfrak{a}_2 \rangle$ be the Lie subalgebra of $\mathfrak{n}$ generated by $\mathfrak{a}_1$ and $\mathfrak{a}_2$ and let $\bar{\mathfrak{a}}_\ell$ be the smallest ideal of $\mathfrak{h}$ containing $\mathfrak{a}_\ell$. Denote by the corresponding capital letters the corresponding Lie groups. So $A_\ell = \exp \mathfrak{a}_\ell = N_{j_\ell}^{S_\ell}$, $B = \exp \mathfrak{h} = \langle A_1 \cup A_2 \rangle$ by 2.5.9, $\bar{A}_\ell = \exp \bar{\mathfrak{a}}_\ell = \langle A_\ell \rangle^B$ by 2.5.10. We have $(\bar{A}_1, \bar{A}_2) = \exp [\bar{\mathfrak{a}}_1, \bar{\mathfrak{a}}_2]$ by 2.5.12.

<u>Assertion</u> $(A_1, A_2) = (\bar{A}_1, \bar{A}_2)$.

First note that (A_1, A_2) is a normal subgroup of $B = \langle A_1 \cup A_2 \rangle$, since if $x_\ell \in A_\ell$, $\ell = 1,2$, and $y \in A_1$ then $(x_1, x_2)^y = (x_1 y, x_2) \cdot (y, x_2)^{-1}$ by 2.2.3,2'), so A_1 normalizes (A_1, A_2) and similarly for A_2 . Let $\pi : B \to B/(A_1, A_2)$ be the natural homomorphism. Then πA_1 is centralized by πA_2 and normalized by πA_1 , hence normal in πB , similarly $\pi A_2 \triangleleft \pi B$. So $(\pi \bar{A}_1, \pi \bar{A}_2) = (\pi A_1, \pi A_2) = \{e\}$, i.e. $(\bar{A}_1, \bar{A}_2) \subset (A_1, A_2)$.

Going back through the definitions we see that the assertion implies

$$N_i^S \cap (N_{j_1}^{S_1}, N_{j_2}^{S_2}) = N_i^S \cap (A_1, A_2) = N_i^S \cap (\bar{A}_1, \bar{A}_2) =$$
$$= \exp(\mathfrak{n}_i^S \cap [\bar{\mathfrak{a}}_1, \bar{\mathfrak{a}}_2]) \supset \exp(\mathfrak{n}_i^S \cap \langle [\mathfrak{n}_{j_2}^{S_1}, \mathfrak{n}_{j_2}^{S_2}] \rangle) \quad \text{if} \quad j_1 + j_2 = i \quad .$$

The claimed inclusion follows now as in the proof of a), where one replaces 4.3.1a) by 4.3.1c) □

Every line L in $\text{Hom}(Q,\mathbb{R})$ contains exactly two open rays, S_+ and S_- say. Define

$$\text{(4.4.8)} \qquad M^L := \langle M^{S_+} \cup M^{S_-} \rangle$$

and inductively

(4.4.9) $$M_1^L := M^L$$

(4.4.10) $$M_i^L := \langle M_i^{S_+} \cup M_i^{S_-} \cup \bigcup (M_j^L, M_k^L) \; ; \; j+k=i \rangle \quad .$$

Note that we have little information about the groups M_i^L - they are generated by iterated commutators - , whereas we know the groups M_i^C , $C \in \mathcal{C}$, very well, because they are images under φ^C .

4.4.11 Lemma *a)* M_i^L *,* $i \in \mathbb{N}$ *, is a filtration of* M^L *.*
b) Suppose $C \in \mathcal{C}$ *and* $C + L \subset C$ *. Then*

$$(M_j^C, M_k^L) \subset M_{j+k}^C \quad .$$

The main feature of b) is that the complicated iterated commutators in M_k^L have commutators with M_j^C in something well known, namely in the image of φ^C .

Proof. a) is true by definition 4.4.10.

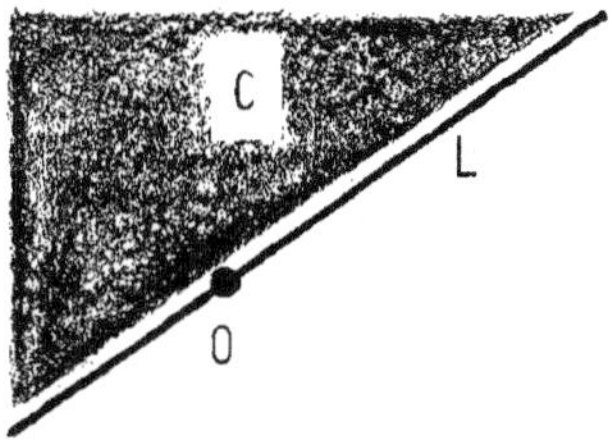

b) Set $A_i = \langle M_i^L \cup M_i^C \rangle$ and $B_i = M_i^C$. By 4.4.5 the sequence $B_1 \supset B_2 \supset \dots$ is a filtration of B_1 and

$$(B_i, M_j^{S_+}) \subset B_{j+i} \quad ,$$
$$(B_i, M_j^{S_-}) \subset B_{j+i} \quad ,$$

if S_+ and S_- are the two open rays contained in L . So B_i is normal in A_1 for every $i \in \mathbb{N}$, by the easy group theoretical lemma 4.4.17a) proved at the end of this section. Define $D_j = \{x \in A_1 \; ; \; (B_i, x) \subset B_{j+i}$ for every $i \in \mathbb{N}\}$. Then $M_j^{S_+} \cup M_j^{S_-} \subset D_j$ and by induction on j we have $A_j \subset D_j$, since $(D_{j_1}, D_{j_2}) \subset D_{j_1+j_2}$ by 4.4.17b) □

Let $\mathcal{L}$ be the set of all lines in $\mathrm{Hom}(Q,\mathbb{R})$.

4.4.12 Proposition *Let* $M = M_1 \supset M_2 \supset \dots$ *be the descending central series of* M *. Suppose any two elements of* R *are positively independ-*

ent. Then

$$M_i = \langle \cup M_i^L \; ; \; L \in \mathcal{L} \rangle \; .$$

PROOF. Set $M^i = \langle \cup M_i^L ; L \in \mathcal{L} \rangle$ for the purpose of this proof. For $i = 1$ we have $M^1 = \langle \cup M^S ; S \in \mathcal{S} \rangle = \langle \cup M^C ; C \in \mathcal{C} \rangle = M = M_1$, the second equation by 4.4.6.

We first show that $(M, M^i) \subset M^{i+1}$ which implies $M_i \subset M^i$ by induction. The sequence $M^1, M^2, \dots$ is descending by 4.4.11a). Let $S \in \mathcal{S}$, $L \in \mathcal{L}$, then either $S \subset L$, hence $(M_i^L, M_j) \subset M_{i+j}^L \subset M^{i+j}$ by 4.4.11a), or else $S + L \in \mathcal{C}$, hence $(M_i^L, M_j^S) \subset M_{i+j}^{S+L} \subset M^{i+j}$ by 4.4.11b) and 4.4.6, thus

$$(M_i^L, M_j^S) \subset M^{i+j}$$

in any case. So for every $i \in \mathbb{N}$ the group M^i is normal in M and $(M^i, M) \subset M^{i+1}$ by 4.4.17a).

The converse inclusion $M^i \subset M_i$ is also proved by induction on i . So suppose $M^j = M_j$ for $j < i$. Note that this implies in particular $M_j^S \subset M_j$ for $j < i$ and $S \in \mathcal{S}$ by definition of M^j , hence $M_j^C \subset M_j$ for $j < i$ and $C \in \mathcal{C}$, by 4.4.6.

Looking at the definition of M_i^L we see that the only thing left to prove is that $M_i^S \subset M_i$. But this follows from 4.4.7a) □

We can and have to do better than 4.4.12. We can exclude an arbitrarily chosen line $L_o \in \mathcal{L}$ from the set of generators of M_i , at the expense of allowing a certain normal subgroup A_i which is trivial for i greather than the step of nilpotency of $\mathfrak{n}$.

4.4.13 PROPOSITION *Suppose any two elements of* R *are positively independent. Set* $A_i = \langle \cup M_i^S ; S \in \mathcal{S} \rangle^M$. *Let* $L_o \in \mathcal{L}$. *Then we have for the i-th term* M_i *of the descending central series of* M

$$M_i = \langle \cup M_i^L , L \in \mathcal{L} , L \neq L_o \rangle \cdot A_i \; .$$

Before proving the proposition let us prove the main result as a corollary.

4.4.14 COROLLARY *If* $\mathfrak{n}$ *is step* s *nilpotent, then* M *is nilpotent of step* $\leq s+1$.

PROOF. We show that M_{s+1} is central in M . It suffices to show that M_{s+1} is centralized by M^S for every $S \in \mathcal{S}$, by 4.4.6. Fix $S \in \mathcal{S}$, let L_o be the line containing S . Then $M_{s+1} = \langle \cup M^L_{s+1}, L \in \mathcal{L}, L \neq L_o \rangle$ by 4.4.13, since $\mathfrak{n}_{s+1} = 0$, hence $A_{s+1} = \{e\}$. Now $(M^L_{s+1}, M^S) \subset M^{L+S}_{s+2}$ for $L \neq L_o$ by 4.4.11b), since $L + S \in \mathcal{C}$, and $M^{L+S}_{s+2} = \{e\}$ by 4.4.4. Therefore $(M^S, M_{s+1}) = \{e\}$ □

4.4.15 REMARK By the corollary the step of M is either s or $s+1$. Both cases occur. The first case occurs e.g. if $N = N^C$ for some $C \in \mathcal{C}$, i.e. if there is a $t \in Q$ such that t acts contracting on N .

An example for the second case is furnished by the group $Q \ltimes N/Z$ of example 3.3.1. In the notations of that example we have isomorphisms $N^C \to (N/Z)^C$ for every $C \in \mathcal{C}$, hence $\varinjlim N^C \cong \varinjlim (N/Z)^C$ and we have a homomorphism $\varinjlim (N/Z)^C \to N$, which is surjective by 4.2.9. The group N/Z is step 2 nilpotent, but N is step 3 nilpotent, therefore $\varinjlim (N/Z)^C = M$ is nilpotent of step ≥ 3 .

PROOF OF PROPOSITION 4.4.13 Fix $L_o \in \mathcal{L}$ for the whole proof. Define $G_i = \langle \cup M^L_i , L \neq L_o \rangle \cdot A_i$ for $i \in \mathbb{N}$. We have $G_i \supset G_{i+1}$ and

$$G_i \subset M_i$$

by 4.4.12. We shall apply 4.4.11 in the following form. For a given ray S and a given line L we have

$$(M^S_j, M^L_i) \subset M^{L+S}_{j+i} \subset G_{j+i} \quad \text{if} \quad S \not\subset L \qquad (1),$$
$$(M^S_j, M^L_i) \subset M^L_{j+i} \quad \text{if} \quad S \subset L \qquad (2).$$

Hence if $S \in \mathcal{S}$ and $L \neq L_o$ we have $(M^S, M^L_i) \subset G_{i+1}$ in any case.

So G_i is normal in G, by 4.4.17a) applied to G/A_i. Define

$$D_j = \{x \in M ; (x,G_i) \subset G_{i+j} \text{ for every } i \in \mathbb{N}\} .$$

Every D_j is a normal subgroup of M and

$$(D_{j_1}, D_{j_2}) \subset D_{j_1+j_2} \tag{3}$$

by 4.4.17b). Obviously $G_1 = M$. We shall prove that $D_1 = M$. This implies $(M,G_i) \subset G_{i+1}$ for every $i \in \mathbb{N}$ and hence $G_i \supset M_i$ by induction, which proves our claim, since $G_i \subset M_i$.

<u>Assertion 1</u> $M_j^S \subset D_j$ if $S \not\subset L_o$, $j \in \mathbb{N}$.

<u>Proof.</u> Let S be an open ray, $S \not\subset L_o$. For every $L \in \mathcal{L}$, including L_o, we have $(M_j^S, M_i^L) \subset G_{i+j}$, by (1) if $S \not\subset L$ and by (2) if $S \subset L$, since in this case $L \neq L_o$. So the image of M_j^S in M/G_{i+j} centralizes a set of generators of M_i/G_{i+j}, hence $(M_j^S, M_i) \subset G_{i+j}$, in particular $(M_j^S, G_i) \subset G_{i+j}$.

<u>Assertion 2</u> $M_j^L \subset D_j$ for every line $L \neq L_o$, $j \in \mathbb{N}$.

This follows from assertion 1 by the definition of M_j^L and (3).

<u>Assertion 3</u> If S_o is an open ray contained in L_o and $S_o \cap R = \emptyset$, then

$$M_j^{S_o} \subset D_j .$$

This follows from assertion 1, 4.4.7b) and (3).

<u>Assertion 4</u> $D_1 = M$.

<u>Proof.</u> Looking at the set $\cup M^S$, $S \in \mathcal{S}$, of generators of M we see that $D_1 = M$ by assertion 1 and assertion 3 unless there is a ray $S_o \subset L_o$ such that $S_o \cap R \neq \emptyset$. So suppose $S_o \in \mathcal{S}$, $S_o \subset L_o$ and $S_o \cap R \neq \emptyset$. Then the opposite ray $-S_o$ does not intersect R, since any two elements of R are positively independent: $-S_o \cap R = \emptyset$. Then we have

$(M^S, G_i) \subset G_{i+1}$ if $S \neq S_o$ by assertion 1 and 3,
$(M^{S_o}, M_i^{S_o}) \subset G_{i+1}$ clearly,
$(M^{S_o}, M_i^L) \subset G_{i+1}$ if $S_o \not\subset L$, by assertion 2,

$(M^{S_o}, M_i^{-S_o}) \subset G_{i+1}$ by assertion 3, since $-S_o \cap R = \emptyset$.
Let E_i be the subgroup of M generated by $M_i^{S_o} \cup M_i^{-S_o} \cup \bigcup M^L$, $L \neq L_o$. Note that $\langle E_i \rangle^M = G_i$. Let $\pi : M \to M/G_{i+1}$ be the natural homomorphism. By the computations above the centralizer of πE_i contains the set $\cup \pi M^S$, $S \in \mathcal{S}$, of generators of πM , so πE_i is central hence normal in πM and $(\pi M, \pi E_i) = \{e\}$. Therefore $E_i \cdot G_{i+1}$ is a normal subgroup of M containing E_i and contained in G_i , hence $E_i \cdot G_{i+1} = G_i$. Finally $(M, G_i) = (M, E_i \cdot G_{i+1}) \subset G_{i+1}$, i.e. $M = D_1$ □

Later on we shall need the following corollary of 4.4.12.

4.4.16 COROLLARY *$\pi : M \to N$ induces an isomorphism*

$$\pi^{ab} : M^{ab} \to N^{ab} .$$

PROOF. For every open ray $S \in \mathcal{S}$ we have a homomorphism $\varphi^S : N^S \to M^S$. Since $\varphi(N_2^S) = M_2^S \subset M_2$ by 4.4.12, φ^S induces a homomorphism $\overline{\varphi^S} : N^S/N_2^S \to M^{ab}$. Take the direct sum of the maps $\overline{\varphi^S}$ over all $S \in \mathcal{S}$. The map $\oplus \overline{\varphi^S} : \oplus N^S/N_2^S = N^{ab} \to M^{ab}$ is inverse to π^{ab} , since $\cup N^S$, $S \in \mathcal{S}$, generates N and $\cup M^S$, $S \in \mathcal{S}$, generates M □

We have used the following easy group theoretical lemma.

4.4.17 LEMMA *a) Let $A \supset B$ be subgroups of a group G . Suppose $G = \langle X \rangle$, $A = \langle Y \rangle$. If*

$$(x,y) \in B \quad \textit{and} \quad (x^{-1},y) \in B$$

for every $x \in X$ and $y \in Y$, then A is a normal subgroup of G and

$$(A,G) \subset \langle B \rangle^G .$$

b) Let $G_1 \supset G_2 \supset \ldots$ be a descending sequence of normal subgroups of G . For $j \in \mathbb{N}$ define

$$D_j = \{x \in G ; (x, G_i) \subset G_{i+j} \text{ for every } i \in \mathbb{N}\} .$$

Then every D_j is a normal subgroup of G and

$$(D_{j_1}, D_{j_2}) \subset D_{j_1+j_2} .$$

PROOF. a) For every $x \in X$ the set

$$A \cap x A x^{-1} = \{y \in A ; (x,y) \in A\}$$

is a group containing Y, hence

$$x A x^{-1} \supset A$$

and similarly $x^{-1} A x \supset A$. So the normalizer N of A is a group containing X, hence $N = G$. So A is normal in G. Factoring out the normal subgroup $\langle B\rangle^G$ of G we see that the centralizer $C(y)$ of every $y \in Y$ contains X, hence $C(y) = G$. So the center Z of G contains Y, hence $A \supset Z$.

b) D_j is the intersection of the kernels of the action of G on G_i/G_{i+j} induced by conjugation, hence normal. The second claim follows from 2.3.1 □

4.5 H CHANGED INTO A TOPOLOGICAL GROUP H_U

Even it $Q \ltimes N$ has a compact presentation, the map $\pi : Q \ltimes M \to M \ltimes N$ need not be an isomorphism. This will be shown later on by an example (see 5.7.4). One has to take into account the topology of N. Therefore we introduce further relations in H to make it into a topological group locally isomorphic to $Q \ltimes N$.

Let U be an open compact subgroup of N. Every compact subset of N is contained in an open compact subgroup of N by 2.6.3. Let H_U be the colimit of the following ordered set of subgroups of $Q \ltimes N$: $\{Q \ltimes N^C, C \in \mathcal{C}\} \cup \{(Q \ltimes N^C) \cap U ; C \in \mathcal{C}\} \cup \{U\}$. So H_U is also the free product of the set of groups $\{Q \ltimes N^C ; C \in \mathcal{C}\} \cup \{U\}$ amalgamated along their intersections

(4.5.1) $$H_U = \coprod_{\cap}(Q \ltimes N^C) \coprod_{\cap} U .$$

We thus have a map $\Psi : \bigcup(Q \ltimes N^C) \cup U \to H_U$ such that $\Psi^C := \Psi | Q \ltimes N^C : Q \ltimes N^C \to H_U$ and $\Psi^U := \Psi | U : U \to H_U$ are group homomorphisms. H_U can also be decribed as a group universal with respect to the existence of such a map.

There is a unique group homomorphism

(4.5.2) $$\pi_U : H_U \to Q \ltimes N$$

such that $\pi_U \circ \Psi$ is the inclusion of $\bigcup(Q \ltimes N^C) \cup U$ into $Q \ltimes N$. Also, there is a unique homomorphism

(4.5.3) $$\rho : H \to H_U$$

such that $\rho \circ \varphi = \Psi | \bigcup(Q \ltimes N^C)$, where φ is as in 4.2.8. We have

(4.5.4) $$\pi_U \circ \rho = \pi .$$

We shall show the following pack of claims.

4.5.5 Lemma *a) Suppose* $0 \notin R$ *. If* U *is an open subgroup of* N*, then* $\langle \bigcup(U \cap N^C), C \in \mathcal{C} \rangle$ *is an open subgroup of* N*.*
b) If U *is an open subgroup of* N *generated by* $\langle \bigcup(U \cap N^C), C \in \mathcal{C} \rangle$*, then* H_U *is generated by* $Q \cup \bigcup(U \cap N^C)$*,* $C \in \mathcal{C}$*. In particular,* $\rho : H \to H_U$ *is surjective in this case.*

4.5.6 Lemma *Suppose* $0 \notin R$*. Let* U *be an open subgroup of* N*. The group* H_U *has a unique topology such that* H_U *is a topological group and* $\Psi^U : U \to H_U$ *is a topological isomorphism of* U *onto an open subgroup of* H_U*.*

4.5.7 Proposition *Suppose* $0 \notin R$*. Let* U *be an open compact subgroup of* N*. The locally compact topological group* H_U *has a compact presentation.*

Proof of 4.5.5 a) We may assume that U is compact, since every compact neighborhood of e generates a subgroup which is open and compact, by 2.6.3. Let V be the subgroup of U generated by $\bigcup(U \cap N^C)$, $C \in \mathcal{C}$. By our hypothesis $0 \notin R$, the image of V in $N^{ab} \simeq n^{ab}$ is a subgroup containing a neighborhood of zero, hence is open, so V is open by 2.6.2 a).

b) H_U is generated by $U \cup \bigcup(Q \ltimes N^C)$, $C \in \mathcal{C}$. The group $Q \ltimes N^C$ is generated by $Q \cup \bigcup(U \cap N^C)$, since there is a contracting element $t \in Q$ for N^C, by 4.1.3. This implies b) □

PROOF OF 4.5.6 Uniqueness is clear, since translations in a topological group are homeomorphisms. To prove existence define $\mathcal{V} = \{\Psi(V)$; V a neighborhood of e contained in $U\}$. Then there is a topology on H_U such that H_U is a topological group and $\mathcal{V}$ is a neighborhood base of e in H_U iff the following conditions hold. 1) For every $\tilde{V} \in \mathcal{V}$ there is a $\tilde{W} \in \mathcal{V}$ such that $\tilde{W} \cdot \tilde{W}^{-1} \subset \tilde{V}$. 2) For every $\tilde{V} \in \mathcal{V}$ and $x \in H_U$ there is a $\tilde{W} \in \mathcal{V}$ such that $x\tilde{W}x^{-1} \subset \tilde{V}$ (see [23] Topologie Gênérale III. § 1. N° 2. Proposition 1). Since $\Psi : U \to H_U$ is a homomorphism, the first condition is satisfied and the second one is satisfied for $x \in \Psi(U)$. Since it suffices to show the second condition for x and x^{-1} in a set of generators of H_U, it remains to prove it for $x \in \Psi(Q)$. Let $V \subset U$ be a neighborhood of e and let $t \in Q$. We show that there is an open subgroup W of U such that $\Psi(t)\Psi(W)\Psi(t)^{-1} \subset \Psi(V)$. We may assume that V is an open compact subgroup of U, by 2.6.1 or [23] Top. Gên. III. N° 6 Cor. 1 de Prop. 14. Define for $C \in \mathcal{C}$ the subgroup $L^C = V \cap t^{-1} V t \cap N^C$. Since $\Psi^C : Q \ltimes N^C \to H_U$ is a homomorphism, we have $\Psi(t)\Psi(L^C)\Psi(t^{-1}) = \Psi(tL^Ct^{-1}) \subset \Psi(V)$. So the subgroup W of U generated by $\cup L^C = \cup(V \cap t^{-1} V t \cap N^C)$ is contained in the group $\{u \in U ; \Psi(u) \in \Psi(t^{-1})\Psi(V)\Psi(t)\}$ and is open by lemma 4.5.5 a).

So we have a topology on H_U for which H_U is a topological group, $\Psi : U \to H_U$ is continuous and $\pi_U : H_U \to Q \ltimes N$ is continuous at e, hence everywhere. The image $\Psi(U)$ is a subgroup of H_U containing a neighborhood of e , hence is open and $\Psi : U \to \Psi(U)$ and $\pi_U : \Psi(U) \to U$ are continuous homomorphisms which are inverse of each other □

PROOF OF 4.5.7 The group H_U is by definition the colimit of the ordered set of groups $\{Q \ltimes N^C, C \in \mathcal{C}\} \cup \{(Q \ltimes N^C) \cap U ; C \in \mathcal{C}\} \cup \{U\}$ which is finite and all groups involved have a compact presentation, by 4.1.3 for $Q \ltimes N^C$ and by 1.1.2 for the compact groups. So 4.2.6 applies □

4.5.8 Note that the definition of H_U can easily be rewritten as an explicit compact presentation, making use of 4.2.6, the explicit presentation in the proof 1.3.1 and the fact that for every compact group L the pair $< L$; all relations of length $\leq 3 >$ is a presentation.

4.6 THE KERNEL OF $H_U \to Q \ltimes N$

Let U be an open compact subgroup of N generated by $U(U \cap N^C)$, $C \in \mathcal{C}$. Such groups exist by 4.5.5 a) if $0 \notin R$. The main result of this section is that if any two elements of R are positively independent then either $\pi_U : H_U \to Q \ltimes N$ is an isomorphism (and the definition of H_U gives an explicit compact presentation of $Q \ltimes N$, see 4.5.8) or $Q \ltimes N$ has no compact presentation at all. Furthermore we show that $\ker \pi_U$ is central in $M_U := \pi_U^{-1}(N)$ and every element of $\ker \pi_U$ has a finite conjugacy class in H_U .

We shall make use of the notions and results of 2.4.

4.6.1 PROPOSITION *Suppose any two elements of* R *are positively independent. Let* U *be an open compact subgroup of* N *generated by* $U(U \cup N^C)$, $C \in \mathcal{C}$. *Let* K_U *be the kernel of* $\pi_U : H_U \to Q \ltimes N$ *and define* $M_U = \pi_U^{-1}(N)$. *Then we have*

a) M_U *is a radicable nilpotent group and is the p-isolator of any of its open subgroups.*

b) K_U *is a discrete central radicable p-torsion subgroup of* M_U *. The group* K_U *is the torsiön subgroup of* M_U *and is contained in* M_U' *.*

Here we denote by p the characteristic of the residue field of K, i.e. $p = \mathrm{char}(o/m)$, where $o = \{x \in K ; \mathrm{mod}_K(x) \leq 1\}$ and $m = \{x \in K ; \mathrm{mod}_K(x) < 1\}$, see 1.2, 2.6.

PROOF. a) The homomorphisms $\pi_U : H_U \to Q \ltimes N$ and $\rho : H \to H_U$ are surjective, the second one by 4.5.5 b). So we have surjective homomorphisms $M \twoheadrightarrow M_U \twoheadrightarrow N$. Hence M_U is nilpotent, since M is nilpotent by 4.4.14. Furthermore the composition of $M^{ab} \twoheadrightarrow (M_U)^{ab} \twoheadrightarrow N^{ab}$ is an isomorphism by 4.4.16, hence $M^{ab} \tilde{\to} (M_U)^{ab} \tilde{\to} N^{ab}$. If follows that $K_U \subset M_U'$. Now $N^{ab} \simeq n^{ab}$ is radicable and is the p-isolator of any of its open subgroups. So the same holds for M_U by 2.4.2 and 2.4.3.
b) The intersection of K_U with the open subgroup $\Psi(U)$ of H_U is $\{e\}$, since $\pi_U|\Psi(U)$ is injective. Therefore $\{e\}$ is an isolated point of K_U, hence K_U is discrete. Furthermore for $x \in K_U$ there is a power p^n of p such that $x^{p^n} \in K_U \cap \Psi(U) = \{e\}$, since M_U is the p-isolator of $\Psi(U)$, hence K_U is a p-torsion group. If some power x^m of an element $x \in M_U$ belongs to K_U, so does x, since N is torsion free. This implies that K_U is radicable, since M_U is radicable. It also implies that every torsion element of M_U belongs to K_U. The remaining claim, that K_U is central, follows from 2.4.5 □

4.6.2 COROLLARY *Either* $\pi_U : H_U \to Q \ltimes N$ *is an isomorphism - and hence* $Q \ltimes N$ *has a compact presentation - or* $Q \ltimes N$ *has no compact presentation.*

Note that the corollary holds for *every* U as in 4.6.1.

PROOF. H_U has a compact presentation by 4.5.7. In order to finish the proof we show that $Q \ltimes N$ has no compact presentation if π_U is not an isomorphism. Define $K_U^n = \{x \in K_U ; x^{p^n} = e\}$ for $n \geq 0$. Every

K_U^n is a normal subgroup of H_U, since it is a characteristic subgroup of K_U. Let $\mu : K_U \to K_U$ be the homomorphism $\mu(x) = x^p$. We have $\mu^{-1}K_U^{n-1} = K_U^n$, by definition, and $\mu(K_U^n) = K_U^{n-1}$, since K_U is p-radicable. So μ induces isomorphisms $K_U^{n+1}/K_U^n \cong K_U^n/K_U^{n-1}$ for $n \geq 1$. We have $K_U = \cup K_U^n$, since K_U is p-torsion. Hence $K_U^{n+1} \neq K_U^n$ if $K_U \neq \{e\}$. If $Q \ltimes N$ had a compact presentation, the discrete kernel K_U of the open surjection $\pi_U : H_U \to Q \ltimes N$ would be finitely generated as a normal subgroup of H_U, by 1.1.3 b), hence contained in some K_U^n, which is impossible unless $K_U = \{e\}$ □

The results of the rest of section 4.6 are not needed elsewhere. We show that the groups K_U^n are actually finite. To do this we regard $\operatorname{gr} M_U$ as a topological Lie algebra.

4.6.3 LEMMA *Let* $M_{U,i}$, N_i, U_i and V_i, $i \in \mathbb{N}$, *be the descending central series of* M_U, N, U and $V = \psi^U(U)$ *resp. Then* U_i *is open in* N_i, $V_i \subset M_{U,i}$ *is open in* $\pi_U^{-1}(N_i)$ *and hence closed in* M *and* $\pi_U : M_{U,i} \to N_i$ *is an open map.*

PROOF. U_i is open in N_i by 2.6.2 b). So $V_i = \psi(U_i)$ is an open subgroup of the open subgroup $\psi(U \cap N_i) = V \cap \pi^{-1}(N_i)$ of $\pi^{-1}(N_i)$. Finally π induces a topological isomorphism of the open subgroup V_i of $M_{U,i}$ onto the open subgroup U_i of N_i □

4.6.4 Let $\operatorname{gr} M_U$ be the associated graded Lie algebra of the descending central series of M_U. It is a locally compact Hausdorff topological Lie algebra. The open continuous surjective homomorphism $\pi_U : M_U \to N$ induces an open continuous surjective homomorphism $\operatorname{gr} M_U \to \operatorname{gr} N$ of Lie algebras. The map $\operatorname{gr}_1 M_U \to \operatorname{gr}_1 M$ is an isomorphism of topological groups (see proof of 4.6.1 a). But $\operatorname{gr}_1 M \simeq \operatorname{gr}_1 \mathfrak{n}$ is a topological $\mathbb{Q}_p$-vector space. It is actually a topological vector space over K, but at the moment we shall make use only of the $\mathbb{Q}_p$-vector space-structure. So we can turn $\operatorname{gr}_1 M_U$ into a topological vector space over $\mathbb{Q}_p$, actually in a unique way. Every k-fold (or perhaps rather k-1-fold) iterated Lie bracket defines a map $gr_1 M_U \times \ldots \times gr_1 M_U \to gr_k M_U$, which is k-multilinear over $\mathbb{Z}$ and continuous and hence induces a continuous homomorphism $gr_1 M_U \otimes_{\mathbb{Q}_p} \ldots \otimes_{\mathbb{Q}_p} gr_1 M_U \to gr_k M_U$ of topological groups. Summing over the various k-fold iterated Lie brackets gives a continuous surjective homomorphism of topological groups $V \to gr_k M_U$, where V is a finite dimensional topological vector space over $\mathbb{Q}_p$. We want to show that $\{x \in gr_k M_U ; p^n x = 0\}$ is finite for every k and n. This is implied by the following lemma, see 4.6.6.

4.6.5 LEMMA *Let* V *be a finite dimensional topological vector space over* $\mathbb{Q}_p$. *Let* W *be a closed subgroup. Then* W/pW *is finite, where* $pW = \{p \cdot w ; w \in W\}$.

PROOF. $W_1 = \{w \in W ; \mathbb{Q}_p w \subset W\}$ is the largest $\mathbb{Q}_p$-vector subspace contained in W. We may assume that $W_1 = 0$, otherwise look at V/W_1. I claim that W is compact. Let $\|\cdot\|$ be a norm for V. If W is not compact, there is a sequence of elements $w_i \in W$ such that $\|w_i\| \to \infty$. There are integers $n_i = p^{r_i}$, $r_i \to \infty$, such that the sequence $n_i w_i$ has a cluster point $w \neq 0$ in W. We may assume that the sequence $n_i w_i$ converges to w. For $s \in \mathbb{Q}_p$ almost all numbers $n_i \cdot s$ are in $\mathbb{Z}_p$, hence $(n_i s) w_i \in \overline{\mathbb{Z} w_i} \subset W$, hence $\lim n_i s w_i = sw \in W$. So W contains the line $\mathbb{Q}_p w$, a contradiction. We may assume furthermore that W spans V. If B is a basis of V contained in W, then $\Sigma \mathbb{Z}_p b$, $b \in B$, is contained in W, so W is open. Then pW is open compact, too, hence W/pW is finite, actually an elementary abelian p-group □

4.6.6 PROPOSITION *Under the hypothesis of 4.6.1 the group*

$$K_U^n = \{x \in K_U ; x^{p^n} = e\}$$

is finite for every $n \in \mathbb{N}$.

PROOF. The argument preceding lemma 4.6.5 shows that $gr_i M_U$ is isomorphic to V/W , where V and W are as in 4.6.5. So the subgroup of elements of $gr_i M_U$ of order dividing p^n is isomorphic to $p^{-n} W/W$, hence finite by 4.6.5. Define $L_i = M_{U,i} \cap K_U^n$. Then L_i/L_{i+1} is isomorphic to a subgroup of $\{x \in gr_i M_U ; p^n x = 0\}$ hence finite. So $L_1 = K_U^n$ is finite □

4.6.7 COROLLARY *Every element of* K_U *has a finite conjugacy class in* H_U .

PROOF. 4.6.6 gives a filtration of K_U by finite characteristic subgroups □

4.7 THE COLIMIT OF THE LIE ALGEBRAS $\mathfrak{n}^C$

We prove in this section the much easier Lie algebra analogues of the preceding results for groups. We shall find that, under our standard hypotheses for this chapter, the kernel of $\varinjlim \mathfrak{n}^C \to \mathfrak{n}$ is central and we obtain information about the action of Q on the kernel. We shall consider fields k of coefficients between $\mathbb{Q}$ and K .

This section has two purposes. On one hand the results will be used later on to give another description of $H_2(\mathfrak{n})^o$ in 5.7.2 and to do the computations for a counterexample in 5.7.4. On the other hand the proofs of this section are an easier counterpart of the computations of 4.4, so these computations here may make it easier to follow those. Actually, the Lie algebra case was proved first and served me as a motivation, guideline and encouragement to solve the group case.

The assumptions and notations of 4.1 and 4.2 will be supposed. Let k be a field with $\mathbb{Q} \subset k \subset K$. If $\mathfrak{a}$ is a Lie algebra over K , we denote by $\mathfrak{a}|k$ the Lie algebra $\mathfrak{a}$ with scalars restricted to k .

Let $\mathfrak{m} := \varinjlim(\mathfrak{n}^C|k)$ be the colimit of the functor $C \longrightarrow \mathfrak{n}^C|k$, $C \in \mathcal{C}$, in the category of Lie algebras over k. The inclusions $\mathfrak{n}^C \to \mathfrak{n}$ induce a homomorphism $\mathfrak{m} \to \mathfrak{n}|k$ of Lie algebras over k.

4.7.1 PROPOSITION *If any two elements of* R *are positively independent, then the kernel of* $\mathfrak{m} \to \mathfrak{n}$ *is central, contained in* $\mathfrak{n}'$ *and contained in the inverse image of* $\mathfrak{n}^0$.

PROOF. Let $\phi : \cup\mathfrak{n}^C \to \mathfrak{m}$ be the map whose restriction to $\mathfrak{n}^C$ is the structure map $\phi^C : \mathfrak{n}^C \to \mathfrak{m}$ for every $C \in \mathcal{C}$. The action of Q on $\mathfrak{n}$ induces an action of Q on $\mathfrak{m}$ by Lie algebra automorphisms over k. The natural map $\tau : \mathfrak{m} \to \mathfrak{n}$ is a map of $k[Q]$-modules. Define $\mathfrak{m}^\beta$ for $\beta \in \mathrm{Hom}(Q,\mathbb{R})$ by $\mathfrak{m}^\beta = \phi\mathfrak{n}^\beta$ for $\beta \neq 0$ and then

$$\mathfrak{m}^0 = \Sigma[\mathfrak{m}^\alpha, \mathfrak{m}^{-\alpha}] \ , \quad 0 \neq \alpha \in \mathrm{Hom}(Q,\mathbb{R}) \quad .$$

4.7.2 LEMMA *Under the hypotheses of 4.7.1 we have*

a) $\mathfrak{m} = \oplus\, \mathfrak{m}^\alpha$, $\alpha \in \mathrm{Hom}(Q,\mathbb{R})$

b) For $\alpha,\beta \in \mathrm{Hom}(Q,\mathbb{R})$ *we have*

$$[\mathfrak{m}^\alpha,\mathfrak{m}^\beta] \subset \mathfrak{m}^{\alpha+\beta}$$

Let us see how the lemma implies the proposition. The map $\tau : \mathfrak{m} \to \mathfrak{n}$ induces an isomorphism $\mathfrak{m}^\beta \to \mathfrak{n}^\beta$ for $\beta \neq 0$, hence $\ker\tau \subset \mathfrak{m}^0 = \tau^{-1}(\mathfrak{n}^0)$. By definition $\mathfrak{m}^0 \subset \mathfrak{m}'$. The ideal $\ker\tau$ is central in $\mathfrak{m}$, since $[\ker\tau, \mathfrak{m}^\beta] \subset \ker\tau \cap \mathfrak{m}^\beta = 0$ for $\beta \neq 0$ and $\Sigma\,\mathfrak{m}^\beta$, $\beta \neq 0$, generates the Lie algebra $\mathfrak{m}$.

PROOF OF THE LEMMA. We first show b), case by case.

1) α and β are positively independent. Then $\mathbb{N}\alpha + \mathbb{N}\beta = C \in \mathcal{C}$, so b) follows by applying ϕ^C in this case.

2) $\alpha \neq 0$, $\beta \neq 0$, α and β positively dependent. The case $\alpha + \beta = 0$ is settled by definition. Suppose $\alpha + \beta \neq 0$. At most one of the rays $\mathbb{R}_+^*\alpha$ or $\mathbb{R}_+^*\beta$ intersects R, say $\mathbb{R}_+^*\alpha \cap R = \emptyset$. Then $\mathfrak{n}^\alpha = \mathfrak{n}_2^\alpha = \Sigma[\mathfrak{n}^\gamma,\mathfrak{n}^\delta]$, γ,δ not in $\mathbb{R}\alpha$, $\gamma + \delta = \alpha$, by 4.3.1 b). Making repeated use of case 1)

and the Jacobi identity we obtain $[m^\alpha, m^\beta] \subset \Sigma[[m^\gamma, m^\delta], m^\beta] \subset$
$\subset \Sigma[m^\gamma, [m^\delta, m^\beta]] + \Sigma[m^\delta, [m^\gamma, m^\beta]] \subset \Sigma[m^\gamma, m^{\delta+\beta}] + \Sigma[m^\delta, m^{\gamma+\beta}] \subset$
$\subset m^{\gamma+\delta+\beta} = m^{\alpha+\beta}$.

3) Exactly one of the two elements α, β is zero. Say $\beta = 0$, $\alpha \neq 0$. Then we have to show that $[m^\alpha, [m^\gamma, m^{-\gamma}]] \subset m^\alpha$ for $\gamma \neq 0$, which follows from 1) and 2) by the Jacobi identity, unless $\alpha + \gamma = 0$ or $\alpha - \gamma = 0$. In the remaining cases at least one of the two rays $\mathbb{R}^*_+\gamma$ or $\mathbb{R}^*_+(-\gamma)$ does not intersect R , say $\mathbb{R}^*_+\gamma \cap R = 0$. By 4.3.1 b) we have $\phi n^\gamma = \phi n_2^\gamma = \Sigma\, \phi[n^\delta, n^\varepsilon] = \Sigma[m^\delta, m^\varepsilon]$ with δ, ε not $\mathbb{R}\gamma$ and $\delta + \varepsilon = \gamma$. Therefore $[m^\gamma, m^{-\gamma}] \subset \Sigma[m^\delta, [m^\varepsilon, m^{-\gamma}]] + \Sigma[m^\varepsilon, [m^\delta, m^{-\gamma}]] \subset$
$\subset \Sigma[m^\delta, m^{\varepsilon-\gamma}] + \Sigma[m^\varepsilon, m^{\delta-\gamma}]$ by cases 1) and 2) and the Jacobi identity. The summands on the right hand side are of the form $[m^\xi, m^{-\xi}]$ with $\xi \notin \mathbb{R}\gamma = \mathbb{R}\alpha$, so we can apply the case already settled to obtain $[m^\alpha, [m^\gamma, m^{-\gamma}]] \subset m^\alpha$ in general.

4) $\alpha = \beta = 0$. Here b) follows immediately from the definition of m^0 , case 3) and the Jacobi identity.

Having proved b) we prove a). The sum on the right hand side is direct, since $\pi | m^\alpha \to n^\beta$ is bijective for $\beta \neq 0$ and the sum of the n^β , $\beta \in \mathrm{Hom}(Q, \mathbb{R})$ is direct. The set of elements $X \in m$ such that $\mathrm{ad}(X)$ maps the vector space $\Sigma\, m^\alpha \subset m$ into itself, as a Lie subalgebra containing the set of generators $\Sigma\, m^\beta$, $\beta \neq 0$, of m , by b). Therefore $\Sigma\, m^\alpha$ is an ideal of m containing a set of generators of m , hence equals m □

4.7.3 REMARK Suppose any two elements of R are positively independent. Set $m = \varinjlim n^C | \mathbb{Q}$. The natural homomorphisms $N^C \to \exp m$, $C \in \mathcal{C}$, induce a group homomorphism $M \to \exp m$. This homomorphism is surjective by 4.7.1, 2.5.7, 4.4.16 and 2.3.8. I do not know whether it is an isomorphism. I do not even know if the kernel of $M \to N$ is central, except in case G is split solvable algebraic over $\mathbb{Q}$ (see 6.2).

V. The second homology

In this chapter we finish the proof of our main theorem 5.6.1. It says that compact presentability of $Q \ltimes N$ is equivalent to the following two conditions. 1) Any two elements of R are positively independent. 2) $H_2(\mathfrak{n})^o = 0$. Here the upper index zero denotes the subspace of vectors of $H_2(\mathfrak{n})$ having relatively compact orbits with respect to the action of Q on $H_2(\mathfrak{n})$. Necessity of the first condition was shown in chapter III. Necessity of the second condition is easy to see (see 5.2.5). Concerning sufficiency we have shown in chapter IV that if the first condition holds, there is a certain extension $\pi_U : H_U \to Q \ltimes N$ which is an isomorphism iff $Q \ltimes N$ has a compact presentation. So the task is to translate the condition $\ker \pi_U = \{e\}$ into the condition $H_2(\mathfrak{n})^o = 0$, which is much easier to check. Before being able to do this in 5.6, we have to discuss several topics, namely group homology in 5.1, Lie algebra homology in 5.2, an isomorphism theorem for group homology and Lie algebra homology with ground field $\mathbb{Q}$ for the second homology in 5.3, a topology on the Hopf extension in 5.4, which will allow us to pass from the second Lie algebra homology with ground field $\mathbb{Q}$ to that over $\mathbb{Q}_p$ and finally to K, by an algebraic proposition in 5.5. In 5.7 we give two examples. We give three descriptions of $H_2(\mathfrak{n})^o$, two in 5.2 and one in 5.7.2.

5.1 Homology of a group. The Hopf extension

In this section we recall the Hopf description of the second homology of a group and define for every group G a central extension of G' with kernel $H_2(G)$, which we call the Hopf extension. It is associated to G in a functorial way. We also show that $H_i(N;\mathbb{Z}) \simeq H_i(N;\mathbb{Q})$ for every $i > 0$ and every nilpotent Lie group N over a field of characteristic zero.

5.1.1 Let G be a group. Let $R \rightarrowtail F \twoheadrightarrow G$ be an exact sequence of groups with F a free group. It induces a central extension

$$E \quad : \qquad R/(F,R) \rightarrowtail F/(F,R) \twoheadrightarrow G$$

of G. This extension depends on the initial exact sequence chosen, but the induced central extension of commutator subgroups

$$E'(G) \quad : \qquad (F' \cap R)/(F,R) \rightarrowtail F'/(F,R) \twoheadrightarrow G'$$

does not, as we shall see now. Here the prime denotes the commutator subgroup $G' = (G,G)$.

5.1.2 Let

$$E_1 \quad : \qquad A \rightarrowtail B \twoheadrightarrow H$$

be a central extension of H, i.e. A is central in B, and let $f : G \to H$ be a group homomorphism. Choose a homomorphism $\varphi : F \to B$ such that the diagram

$$\begin{array}{ccc} F & \twoheadrightarrow & G \\ \downarrow{\scriptstyle\varphi} & & \downarrow{\scriptstyle f} \\ B & \twoheadrightarrow & H \end{array}$$

is commutative. For any two such φ the induced maps $F'/(F,R) \to B$ coincide, as is easily seen, hence f induces a map of exact sequences $E'(G) \to E_1$. In particular, given two exact sequences $R_1 \rightarrowtail F_1 \twoheadrightarrow G_1$ and $R_2 \rightarrowtail F_2 \twoheadrightarrow G_2$ with F_1, F_2 free groups and a group homomorphism $f : G_1 \to G_2$ we have central extensions $R_i/(R_i,F_i) \rightarrowtail F_i/(R_i,F_i) \twoheadrightarrow G_i$

and hence an induced map of central extensions $E'(f) : E'(G_1) \to E'(G_2)$. So E' defines a functor from the category of groups G to the category of isomorphism classes of central extensions of G' , which we call the *Hopf extension*.

5.1.3 Recall that the *homology* $H_*(G;A)$ of the group G with coefficients in a $\mathbb{Z}[G]$-module A is by definition the homology of the group ring $\mathbb{Z}[G]$ with coefficients in the $\mathbb{Z}[G]$-module A . For the trivial $\mathbb{Z}[G]$-module $\mathbb{Z}$ we have

$$H_2(G;\mathbb{Z}) \simeq (F'\cap R)/(F,R)$$

by a theorem of Hopf, where the right hand side is as in 5.1.1, see [32]. Our computations will be done with this description of $H_2(G;\mathbb{Z})$.

We shall show first that $H_2(N;\mathbb{Z})$ is a vector space over $\mathbb{Q}$ in a natural way if $N = \exp \mathfrak{n}$, where $\mathfrak{n}$ is a nilpotent Lie algebra over a field of characteristic zero.

5.1.4 PROPOSITION *Let* $\mathfrak{n}$ *be a nilpotent Lie algebra over* $\mathbb{Q}$. *Let* N *be the corresponding Lie group over* $\mathbb{Q}$ *(see 2.5). Then the coefficient homomorphism* $\mathbb{Z} \to \mathbb{Q}$ *induces an isomorphism*

$$H_i(N;\mathbb{Z}) \xrightarrow{\sim} H_i(N;\mathbb{Q})$$

for $i > 0$. *In particular,* $H_i(N;\mathbb{Z})$ *is a vector space over* $\mathbb{Q}$ *in a natural way for* $i > 0$.

PROOF. Any finitely generated subalgebra of $\mathfrak{n}$ has finite dimension, so $\mathfrak{n}$ is the union of finite-dimensional subalgebras. Since homology commutes with direct limits of groups, we may assume that $\mathfrak{n}$ has finite dimension. We prove our claim by induction on $\dim \mathfrak{n}$. For $\dim \mathfrak{n} = 1$ we have $H_1(\mathbb{Q};\mathbb{Z}) = \mathbb{Q}^{ab} = \mathbb{Q}^{ab} \otimes_{\mathbb{Z}} \mathbb{Q} = H_1(\mathbb{Q};\mathbb{Z}) \otimes_{\mathbb{Z}} \mathbb{Q} = H_1(\mathbb{Q};\mathbb{Q})$, the last equation by the universal coefficient theorem. For $i > 1$ we have $H_i(\mathbb{Q};\mathbb{Z}) = 0 = H_i(\mathbb{Q};\mathbb{Q})$, since $\mathbb{Q}$ is the direct limit of infinite cyclic groups and $H_i(\mathbb{Z};\mathbb{Z}) = 0 = H_i(\mathbb{Z};\mathbb{Q})$. If $\dim \mathfrak{n} > 1$, let $\mathfrak{z}$ be a central one-dimensional ideal of $\mathfrak{n}$. We have a map of spectral sequences from

$$H_i(\exp \mathfrak{n}/\exp \mathfrak{z} \ ; \ H_j(\exp \mathfrak{z} \ ; \ \mathbb{Z})) \Rightarrow H_{i+j}(\exp \mathfrak{n} \ ; \ \mathbb{Z})$$

to

$$H_i(\exp \mathfrak{n}/\exp \mathfrak{z} \ ; \ H_j(\exp \mathfrak{z} \ ; \ \mathbb{Q})) \Rightarrow H_{i+j}(\exp \mathfrak{n} \ ; \ \mathbb{Q})$$

which is an isomorphism on the left hand side if $i + j > 0$. This implies our claim □

5.2 Homology of a Lie algebra

In this section we recall the notion of homology of a Lie algebra, Hopf's description of the second homology and define a Hopf extension for Lie algebras. We also prove necessity of the second condition " $H_2(\mathfrak{n})^o = 0$ " of our main theorem.

5.2.1 Let $\mathfrak{g}$ be a Lie algebra over a field k. Recall that the *Koszul complex* of $\mathfrak{g}$ is the exterior algebra $\Lambda(\mathfrak{g})$ of the vector space $\mathfrak{g}$ over k, regarded as a graded vector space $\Lambda(\mathfrak{g}) = \oplus \Lambda^n(\mathfrak{g})$, together with the k-linear boundary map $d : \Lambda(\mathfrak{g}) \to \Lambda(\mathfrak{g})$ with components $d_n : \Lambda^n(\mathfrak{g}) \to \Lambda^{n-1}(\mathfrak{g})$ defined by

$$d_n(x_1\wedge\cdots\wedge x_n) = \sum_{1\le i<j\le n} (-1)^{i+j}[x_i,x_j]\wedge x_1\wedge\cdots\wedge\hat{x}_i\wedge\cdots\wedge\hat{x}_j\wedge\cdots x_n ,$$

where the roof on x_i means: delete x_i. The *homology* $H_*(\mathfrak{g}) = \oplus H_i(\mathfrak{g})$ of the Lie algebra $\mathfrak{g}$ is by definition the homology of the Koszul complex, i.e.

$$H_i(\mathfrak{g}) = \frac{\ker d_i}{\operatorname{im} d_{i+1}} .$$

So we always take coefficients in the trivial $\mathfrak{g}$-module k.

5.2.2 We have the analogous statements for Lie algebras as in 5.1.2 for groups. All Lie algebras will have ground field k. Let

$$\mathfrak{E}_1 : \qquad \mathfrak{a} \rightarrowtail \mathfrak{b} \twoheadrightarrow \mathfrak{c}$$

be a central extension of $\mathfrak{c}$, i.e. an exact sequence of Lie algebras with $\mathfrak{a}$ central in $\mathfrak{h}$. Let $f : \mathfrak{g} \to \mathfrak{c}$ be a homomorphism of Lie algebras. Let $\mathfrak{f} \twoheadrightarrow \mathfrak{g}$ be a surjective Lie algebra homomorphism with $\mathfrak{f}$ a free Lie algebra. Let $\mathfrak{r}$ be the kernel of $\mathfrak{f} \twoheadrightarrow \mathfrak{g}$. There is a Lie algebra homomorphism $\varphi : \mathfrak{f} \to \mathfrak{h}$ making the diagram

$$\begin{array}{ccc} \mathfrak{f} & \to & \mathfrak{g} \\ \downarrow{\scriptstyle \varphi} & & \downarrow{\scriptstyle f} \\ \mathfrak{h} & \to & \mathfrak{c} \end{array}$$

commutative. For any two such φ the induced maps $\mathfrak{f}'/[\mathfrak{f},\mathfrak{r}] \to \mathfrak{h}$ coincide, hence f induces a map of exact sequences $\mathfrak{E}'(\mathfrak{g}) \to \mathfrak{E}_1$, where

$$\mathfrak{E}'(\mathfrak{g}) : (\mathfrak{f}'\cap\mathfrak{r})/[\mathfrak{f},\mathfrak{r}] \rightarrowtail \mathfrak{f}'/[\mathfrak{f},\mathfrak{r}] \twoheadrightarrow \mathfrak{g}' .$$

As for groups, $\mathfrak{E}'$ is a functor from Lie algebras $\mathfrak{g}$ to central extensions of $\mathfrak{g}'$. We call $\mathfrak{E}'$ the *Hopf extension* functor.

5.2.3 Hopf's description of $H_2(\mathfrak{n})$

Let $\mathfrak{g}$ be a Lie algebra. Let

$$\mathfrak{r} \rightarrowtail \mathfrak{f} \twoheadrightarrow \mathfrak{g}$$

be an exact sequence of Lie algebras with $\mathfrak{f}$ free. Then

$$H_2(\mathfrak{g}) \simeq (\mathfrak{f}'\cap\mathfrak{r})/[\mathfrak{f},\mathfrak{r}]$$

by a theorem of Hopf, see [32]. The Hopf isomorphism $H_2(\mathfrak{g}) \to (\mathfrak{f}'\cap\mathfrak{r})/[\mathfrak{f},\mathfrak{r}]$ is defined as follows. Let $\pi : \mathfrak{f}/[\mathfrak{f},\mathfrak{r}] \twoheadrightarrow \mathfrak{g}$ be induced by $\mathfrak{f} \twoheadrightarrow \mathfrak{g}$. Define

$$\xi : \mathfrak{g} \wedge \mathfrak{g} \to \mathfrak{f}'/[\mathfrak{f},\mathfrak{r}]$$

by

$$\xi(X\wedge Y) = [\pi^{-1}(X),\pi^{-1}(Y)] .$$

ξ is well defined, since the kernel of π is central. ξ vanishes on $\operatorname{im} d_3$, by the Jacobi identity, and maps $\ker d_2$ onto $(\mathfrak{r}\cap\mathfrak{f}')/[\mathfrak{f},\mathfrak{r}]$, clearly, hence induces a map

$$\bar{\xi} : H_2(\mathfrak{g}) \to (\mathfrak{f}'\cap\mathfrak{r})/[\mathfrak{f},\mathfrak{r}] ,$$

the Hopf isomorphism.

5.2.4 Let V be a vector space of finite dimension over a local field K . Suppose a representation of a group G on V is given.

Define V^o as the set of vectors of V, whose G-orbit has compact closure. V^o is a $K[G]$-submodule of V. If G is a free abelian finitely generated group Q, K is a p-adic field and V is a sum of one-dimensional submodules, then this notation is in keeping with our earlier notation (see 3.1.5, although not with 3.1 line 4 for $\lambda = 0$)

$$V^o = \oplus V^\alpha \ , \ \alpha \in \mathrm{Hom}(Q,K^*) \ , \ \nu_*(\alpha) = 0 \ .$$

Here we think of zero as an element of $\mathrm{Hom}(Q,\mathbb{R})$. In particular, suppose $\mathfrak{n}$ is a finite dimensional Lie algebra over a p-adic field K and Q acts by automorphisms on $\mathfrak{n}$. Suppose further that $\mathfrak{n}$ is the sum of one-dimensional submodules, then the same holds for $\Lambda(\mathfrak{n})$ and any $K[Q]$-subquotient thereof, in particular for $H_i(\mathfrak{n})$, Hence

$$H_i(\mathfrak{n})^o = \bigoplus_{\nu_*(\alpha)=0} H_i(\mathfrak{n})^\alpha =$$

$= \{$elements of $H_i(\mathfrak{n})$ whose Q-orbit has compact closure$\}$.

We now assume the hypotheses and notations of 3.1.

5.2.5 PROPOSITION *If* $Q \ltimes N$ *has a compact presentation, then* $H_2(\mathfrak{n})^o = 0$.

PROOF. Let V be a $K[Q]$-submodule of $\mathfrak{n}$ such that the natural map $\mathfrak{n} \to \mathfrak{n}^{ab}$ induces an isomorphism of $K[Q]$-modules $V \xrightarrow{\sim} \mathfrak{n}^{ab}$. Let $\mathfrak{f}$ be the free K-Lie algebra over the vector space V, defined by the following universal property. Every K-linear map from V to any Lie algebra $\mathfrak{g}$ over K extends uniquely to a homomorphism $\mathfrak{f} \to \mathfrak{g}$ of Lie algebra over K. Then $\mathfrak{f}$ is a $K[Q]$-module in a natural way and sum of one-dimensional submodules. The inclusion map $V \to \mathfrak{n}$ induces a Lie algebra homomorphism $\mathfrak{f} \to \mathfrak{n}$ which is a map of $K[Q]$-modules and is surjective by 2.1.3. We thus have an exact sequence

$$\mathfrak{r} \rightarrowtail \mathfrak{f} \twoheadrightarrow \mathfrak{n}$$

with $\mathfrak{r} \subset \mathfrak{f}'$, since $\mathfrak{f}^{ab} = V \xrightarrow{\sim} \mathfrak{n}^{ab}$.

So
$$\mathfrak{r}/[\mathfrak{f},\mathfrak{r}] \rightarrowtail \mathfrak{f}/[\mathfrak{f},\mathfrak{r}] \twoheadrightarrow \mathfrak{n}$$

is a central extension of $\mathfrak{n}$ with $\mathfrak{r}/[\mathfrak{f},\mathfrak{r}] \simeq H_2(\mathfrak{n})$ as $K[Q]$-modules, by naturality of the Hopf isomorphism in 5.2.3. Mod out the $K[Q]$-submodule corresponding to $\oplus H_2(\mathfrak{n})^\alpha$, $\nu_*(\alpha) \neq 0$, to obtain a central extension of $K[Q]$-Lie algebras

$$H_2(\mathfrak{n})^0 \rightarrowtail \mathfrak{g} \twoheadrightarrow \mathfrak{n}$$

with $\mathfrak{g} = (\mathfrak{f}/[\mathfrak{f},\mathfrak{r}])/\bigoplus_{\nu_*(\alpha)\neq 0} H_2(\mathfrak{n})^\alpha$.

Note that $\mathfrak{g}$ has finite dimension, since $H_2(\mathfrak{n})^0$ does by the Koszul-complex description. Take the corresponding Lie groups over K (see 2.5) and form the semidirect product with Q :

$$\exp H_2(\mathfrak{n})^0 \rightarrowtail Q \ltimes \exp \mathfrak{g} \twoheadrightarrow Q \ltimes N \ .$$

If now $Q \ltimes N$ is compactly generated, 3.2.2 yields $0 \notin \nu_*(W(\mathfrak{n}^{ab})) = \nu_*(W(\mathfrak{g}^{ab}))$, since $\mathfrak{g}^{ab} \cong \mathfrak{n}^{ab}$. So $Q \ltimes \exp \mathfrak{g}$ is compactly generated by 3.2.2. Any compact subset of $\exp H_2(\mathfrak{n})^0$ is contained in a compact normal subgroup of $Q \ltimes \exp \mathfrak{g}$ by 5.2.4 and 2.6.3, so the factor group $Q \ltimes N$ is not compactly presentable, unless the K-vector space $H_2(\mathfrak{n})^0$ is zero, by 1.1.3 b) □

5.2.6 In 5.7.2 we shall give another description of $H_2(\mathfrak{n})^0$, namely $H_2(\mathfrak{n})^0$ is isomorphic to the kernel of the map $\varinjlim \mathfrak{n}^C \to \mathfrak{n}$ of 4.7, if any two elements of R are positively independent. This will give another proof of 5.2.5 (see 5.7.5).

5.3 Lie Algebra Homology versus Group Homology

In this section we prove a theorem showing equality of the second homology of a nilpotent Lie algebra and its Lie group (cf. remark 5.3.9). This is an easy application of Malcev correspondence. So we first have to recall the notion of a lattice in a nilpotent Lie group of finite dimension.

5.3.1 Let $\mathfrak{n}$ be a nilpotent Lie algebra over $\mathbb{Q}$. Suppose $\mathfrak{n}$ has finite dimension. A subgroup Γ of the rational Lie group $N = \exp \mathfrak{n}$ over $\mathbb{Q}$ (see 2.5) is called a *lattice* in N , if it is finitely generated and N is the isolator of Γ , i.e. for every $x \in N$ there is an integer $m \neq 0$ such that $x^m \in \Gamma$. Recall that two subgroups Γ_1 , Γ_2 of a group are called *commensurable* if $\Gamma_1 \cap \Gamma_2$ is of finite index in both Γ_1 and Γ_2 . Commensurability is an equivalence relation.

5.3.2 LEMMA *a) The lattices in* N *form a commensurability class.*
b) Let T *be a finite subset of* N . *Then* $\langle T\rangle$ *is a lattice iff the image of* T *in* $N^{ab} \simeq \mathfrak{n}^{ab}$ *spans* $\mathfrak{n}^{ab}$ *as a vector space over* $\mathbb{Q}$.
c) Given a basis B *of the* $\mathbb{Q}$*-vector space* $\mathfrak{n}$ *and a lattice* Γ . *Then there is an integer* $c \neq 0$ *such that*

$$c\Big(\sum_{b\in B} \mathbb{Z}b\Big) \subset \log \Gamma \subset c^{-1}\Big(\sum_{b\in B} \mathbb{Z}b\Big) .$$

PROOF. a) Let Γ_1 and Γ_2 be lattices of N . Then the group Γ generated by $\Gamma_1 \cup \Gamma_2$ is a lattice, by definition, and Γ_1 and Γ_2 are of finite index in Γ , by 2.4.4. Conversely, the commensurability class of a lattice contains only lattices, clearly.

b) follows from 2.4.3.

c) Let T be a finite set of generators of Γ , let $\mathfrak{m}$ be the Lie subring of $\mathfrak{n}$ generated by $\log T$. Then there is a $c \in \mathbb{N}$ and a Lie subring $\mathfrak{h}$ of $\mathfrak{n}$, such that

$$\mathfrak{h} \supset \mathfrak{m} \cup \log \Gamma \supset \mathfrak{m} \cap \log \Gamma \supset c\cdot\mathfrak{h} ,$$

by 2.5.13. Now Γ is the isolator of N , hence there is a number $c_1 \in \mathbb{N}$ such that $c_1 \cdot b \in \log \Gamma$ for every $b \in B$, hence $c_1 \cdot b \in \mathfrak{h}$ for every $b \in B$. On the other hand, $\mathfrak{m}$ is a finitely generated Lie subring of $\mathfrak{n}$, hence a finitely generated abelian subgroup of $\mathfrak{n}$, since it is spanned as a $\mathbb{Z}$-module by the iterated Lie brackets of the elements of $\log T$, of which only a finite number is $\neq 0$, since $\mathfrak{n}$ is nilpotent. So there is a natural number c_2 such that $c_2\cdot\mathfrak{m} \subset \sum \mathbb{Z}b$. We thus have

$$c\cdot c_1\Big(\sum_{b\in B} \mathbb{Z}b\Big) \subset c\,\mathfrak{h} \subset \log \Gamma \subset \mathfrak{h} \subset c^{-1}\mathfrak{m} \subset (c\cdot c_2)^{-1}\Big(\sum_{b\in B} \mathbb{Z}b\Big) \quad \square$$

5.3.3 Every lattice in our sense is a lattice in the usual sense of the word, i.e. is a discrete subgroup of $\exp(\mathfrak{n}\otimes_{\mathbb{Q}}\mathbb{R})$ with compact coset space. More precisely, our lattices Γ in $\exp(\mathfrak{n})$ are exactly the discrete subgroups of $\exp(\mathfrak{n}\otimes_{\mathbb{Q}}\mathbb{R})$ with compact coset space contained in $\exp(\mathfrak{n})$. This follows easily from c) and b) by induction on $\dim \mathfrak{n}$.

5.3.4 Furthermore, the lattices of $\exp \mathfrak{n}$ are precisely the *arithmetic subgroups* (see 6.1) of $\exp \mathfrak{n}$, if $\exp \mathfrak{n}$ is represented as the set of $\mathbb{Q}$-points of a unipotent algebraic group over $\mathbb{Q}$. To see this let $\mathfrak{u}$, U be the Lie algebra (group) of upper triangular $n\times n$-matrices with zeros (ones) on the diagonal and entries $\in \mathbb{Q}$. Then $U = \exp(\mathfrak{u})$, and $U \cap Gl_n(\mathbb{Z})$ is a lattice of U, since its isolator is U by 2.4.3 and it is finitely generated, as is easily seen by induction on n. If $\mathfrak{n}$ is a Lie subalgebra of $\mathfrak{u}$, then $\exp \mathfrak{n} \cap Gl_n(\mathbb{Z})$ has isolator $\exp \mathfrak{n}$ in $\exp \mathfrak{n}$, clearly, and is finitely generated, since every subgroup of a finitely generated nilpotent group is finitely generated, hence $\exp \mathfrak{n} \cap Gl_n(\mathbb{Z})$ is a lattice of $\exp \mathfrak{n}$. An arithmetic subgroup of $\exp \mathfrak{n}$ is by denfinition a group commensurable with $\exp \mathfrak{n} \cap Gl_n(\mathbb{Z})$. So every arithmetic subgroup of $\exp \mathfrak{n}$ is a lattice, by 5.3.2 a), and every lattice is commensurable with our arithmetic group, hence is arithmetic itself.

The following theorem is basic.

5.3.5 THEOREM *(Malcev [24]) For every finitely generated torsion-free nilpotent group* Γ *there is a - unique up to isomorphism - nilpotent Lie algebra* $\mathfrak{n}$ *over* $\mathbb{Q}$ *of finite dimension, such that* Γ *is isomorphic to a lattice of* $\exp \mathfrak{n}$. *For every group homomorphism* $f : \Gamma_1 \to \Gamma_2$ *of lattices* Γ_i *in* $\exp(\mathfrak{n}_i)$, *where* $\mathfrak{n}_i$ *are nilpotent Lie algebras of finite dimension over* $\mathbb{Q}$ *for* $i = 1,2$, *there is a unique homomorphism* $\varphi : \mathfrak{n}_1 \to \mathfrak{n}_2$ *of Lie algebras inducing* f.

5.3.6 We now define our map $\sigma : H_2(N) \to H_2(\mathfrak{n})$. Let $\mathfrak{n}$ be a nilpotent Lie algebra over $\mathbb{Q}$ and let $N = \exp \mathfrak{n}$ be the corresponding Lie group over $\mathbb{Q}$. Note that we do not suppose that $\mathfrak{n}$ is finite-dimensional over $\mathbb{Q}$, in fact, we shall apply this for $\mathfrak{n}$ of infinite dimension. Let $R \rightarrowtail F \twoheadrightarrow N$ and $\mathfrak{r} \rightarrowtail \mathfrak{f} \twoheadrightarrow \mathfrak{n}$ be exact sequences of groups and Lie algebras resp. with F and $\mathfrak{f}$ free. Then $R/(R,F) \rightarrowtail F/(R,F) \twoheadrightarrow N$, $\mathfrak{r}/[\mathfrak{r},\mathfrak{f}] \rightarrowtail \mathfrak{f}/[\mathfrak{r},\mathfrak{f}] \twoheadrightarrow \mathfrak{n}$ are central extensions. In particular, $\mathfrak{f}/[\mathfrak{r},\mathfrak{f}]$ is a nilpotent Lie algebra over $\mathbb{Q}$, so there is a corresponding Lie group $\exp(\mathfrak{f}/[\mathfrak{r},\mathfrak{f}])$ over $\mathbb{Q}$. Let $\varphi : F \to \exp(\mathfrak{f}/[\mathfrak{r},\mathfrak{f}])$ be a group homomorphism making the diagram

$$\begin{array}{ccc} F & \longrightarrow & N \\ \downarrow\varphi & & \| \\ \exp(\mathfrak{f}/[\mathfrak{f},\mathfrak{r}] & \longrightarrow & \exp \mathfrak{n} \end{array}$$

commutative. It induces a homomorphism of exact sequences

$$\begin{array}{ccccc} R/(F,R) & \rightarrowtail & F/(F,R) & \twoheadrightarrow & N \\ \downarrow & & \downarrow & & \| \\ \exp(\mathfrak{r}/[\mathfrak{r},\mathfrak{f}]) & \rightarrowtail & \exp(\mathfrak{f}/[\mathfrak{r},\mathfrak{f}]) & \twoheadrightarrow & \exp \mathfrak{n} \end{array}$$

and hence a homomorphism $E'(N) \to \exp E'(\mathfrak{n})$ of the derived exact sequences, which is independent of the homomorphism φ chosen.

$$\begin{array}{llllll} E'(N) & : & (R\cap F')/(F,R) & \rightarrowtail & F'/(F,R) & \twoheadrightarrow N' \\ & & & & & \quad \| \\ \exp E'(\mathfrak{n}) & : & \exp((\mathfrak{r}\cap\mathfrak{f}')/[\mathfrak{f},\mathfrak{r}]) & \rightarrowtail & \exp(\mathfrak{f}'/[\mathfrak{f},\mathfrak{r}]) & \twoheadrightarrow \exp \mathfrak{n}' . \end{array}$$

The equalities $(\exp \mathfrak{n})' = \exp \mathfrak{n}'$ and $(\exp(\mathfrak{f}/[\mathfrak{r},\mathfrak{f}]))' = \exp(\mathfrak{f}'/[\mathfrak{r},\mathfrak{f}])$ follow from 2.5.7. Composing with the Hopf isomorphisms

$$H_2(N) \cong (R\cap F')/(F,R)$$

and

$$H_2(\mathfrak{n}) \cong (\mathfrak{r}\cap\mathfrak{f}')/[\mathfrak{f},\mathfrak{r}] \cong \exp((\mathfrak{r}\cap\mathfrak{f}')/[\mathfrak{f},\mathfrak{r}])$$

we obtain a map

$$\sigma : H_2(N) \to H_2(\mathfrak{n}) .$$

5.3.7 THEOREM *Let* $\mathfrak{n}$ *be a nilpotent Lie algebra over* $\mathbb{Q}$ *and let* $N = \exp \mathfrak{n}$ *be the corresponding Lie group over* Q. *Then* $\sigma : H_2(N;\mathbb{Z}) \to H_2(\mathfrak{n};\mathbb{Q})$ *is an isomorphism of groups.*

PROOF. Again we suppress the domains $\mathbb{Z}$ and $\mathbb{Q}$ of coefficients from the notation of $H_2(N)$ and $H_2(\mathfrak{n})$ resp.. It suffices to prove the theorem for finite dimensional $\mathfrak{n}$, since $H_2(N) \to H_2(\mathfrak{n})$ is a natural transformation of functors of nilpotent Lie algebras $\mathfrak{n}$ over $\mathbb{Q}$ and everything commutes with direct limits. The theorem will follow from the following lemma. Note that $H_2(N) = H_2(N;\mathbb{Q})$ by 5.1.4.

5.3.8 LEMMA *Let* $\mathfrak{n}$ *be a nilpotent Lie algebra over* $\mathbb{Q}$ *of finite dimension. Let* T *be a subset of* $N = \exp \mathfrak{n}$ *such that* $N \to N^{ab} \simeq \mathfrak{n}^{ab}$ *maps* T *bijectively onto a basis of the* $\mathbb{Q}$*-vector space* $\mathfrak{n}^{ab}$. *Let* Γ *be the subgroup of* N *generated by* T, *a lattice by 5.3.2 b). Then the composed homomorphism*

$$H_2(\Gamma;\mathbb{Q}) \to H_2(N;\mathbb{Q}) \overset{\sigma}{\to} H_2(\mathfrak{n})$$

is an isomorphism, the first homomorphism being induced by the inclusion $\Gamma \to N$.

We first show how this lemma implies the theorem. N is the union of its lattices. More precisely, there is an ascending sequence of lattices of N as in the lemma whose union is N. To see this, let $X_1,\ldots,X_m$ be a finite subset of $\mathfrak{n}$ whose image in $\mathfrak{n}^{ab}$ forms a basis of $\mathfrak{n}^{ab}$ and let n_i, $i \in \mathbb{N}$, be a sequence of non-zero integers such that n_i divides n_{i+1} and every integer divides some n_i. Then every group $\Gamma_i = \langle \exp n_i^{-1} X_j \,,\, j = 1,\ldots,m\rangle$ is a lattice as in the lemma and $\cup\Gamma_i = N$ by 2.3.8 and 2.5.7. Since homology commutes with direct limits, the theorem follows □

PROOF OF LEMMA 5.3.8. Let F be the free group with basis T. Let $F \to \Gamma$ be the homomorphism induced by the inclusion $T \to \Gamma$. Let R be its kernel. The induced map $F^{ab} \to \Gamma^{ab}$ is an isomorphism, since $F^{ab} \to \Gamma^{ab}$ is surjective and the composite map $F^{ab} \to \Gamma^{ab} \to N^{ab}$ is the homomorphism of the free abelian group with basis T into N^{ab}, which is injective by hypothesis. Therefore $R \subset F'$, hence

$$H_2(\Gamma) \rightarrowtail F/(F,R) \twoheadrightarrow \Gamma ,$$

where $H_2(\Gamma) := H_2(\Gamma;\mathbb{Z})$. In particular, every torsion element of $F/(F,R)$ is contained in the central subgroup $H_2(\Gamma)$. Note that $H_2(\Gamma)$ is finitely generated, since it is a subgroup of the finitely generated nilpotent group $F/(F,R)$. Let Δ be $F/(F,R)$ modulo its torsion subgroup. Let $\mathfrak{d}$ be the nilpotent Lie algebra of finite dimension over $\mathbb{Q}$ corresponding to Δ by Malcev's theorem. We have a commutative diagram

$$\begin{array}{ccccc} H_2(\Gamma)/\text{torsion} & \rightarrowtail & \Delta & \twoheadrightarrow & \Gamma \\ \downarrow & & \downarrow & & \downarrow \\ \exp \mathfrak{k} & \xrightarrow{\exp(\phi_1)} & \exp \mathfrak{d} & \xrightarrow{\exp(\phi_2)} & \exp \mathfrak{n} \end{array} ,$$

where $\mathfrak{k}$ is the Lie algebra corresponding to $H_2(\Gamma)/\text{torsion}$ by Malcev correspondence, i.e.

$$\mathfrak{k} = H_2(\Gamma) \otimes_{\mathbb{Z}} \mathbb{Q} ,$$

since $H_2(\Gamma)$ is abelian. The Lie algebra homomorphisms ϕ_i are induced by the homomorphisms of lattices by Malcev correspondence. I claim that the sequence

$$\mathcal{E}_1 : \qquad \mathfrak{k} \xrightarrow{\phi_1} \mathfrak{d} \xrightarrow{\phi_2} \mathfrak{n}$$

is short exact and $\mathfrak{k}$ is central in $\mathfrak{d}$. The $\mathbb{Q}$-linear map ϕ_1 is injective, since it is on a spanning abelian subgroup. The $\mathbb{Q}$-linear map ϕ_2 is surjective, since the image contains the spanning subset $\log \Gamma$. The composition $\phi_2 \circ \phi_1$ is zero, since it is on a spanning subset of $\mathfrak{k}$. Finally, if $x \in \ker \phi_2$, then, for some $m \in \mathbb{N}$, $(\exp x)^m$ belongs to Δ and is mapped to e in Γ , hence $\exp(m \cdot x)$ is in the image of $H_2(\Gamma)$ mod torsion, so x is in the image of $\mathfrak{k}$. In order to see that $\mathfrak{k}$ is central in $\mathfrak{d}$ let $x \in \mathfrak{k}$, $y \in \mathfrak{d}$ and look at the polynomial map $\xi : \mathbb{Q} \times \mathbb{Q} \to \mathfrak{d}$, $(r,s) \to \log(\exp r X, \exp s Y)$. For fixed $r \in \mathbb{Z}$ dividing some integer we have $\exp r X \in H_2(\Gamma)/\text{torsion}$, hence $\xi(s,r) = 0$ for infinitely many $s \in \mathbb{Z}$, hence for every $s \in \mathbb{Q}$, similarly for fixed $s \in \mathbb{Q}$ we have $\xi(s,r) = 0$ for infinitely many $r \in \mathbb{Z}$, hence for every $r \in \mathbb{Q}$. It follows that $[X,Y] = 0$, by 2.5.10.

Now let $\mathfrak{f}$ be the free Lie algebra over $\mathbb{Q}$ with basis T . We have an obvious map of exact sequences $\mathcal{E}(\mathfrak{n}) \to \mathcal{E}_1$,

$E(\mathfrak{n})$: $H_2(\mathfrak{n}) \to \mathfrak{f}/[\mathfrak{f},\mathfrak{r}] \twoheadrightarrow \mathfrak{n}$

E_1 : $\mathfrak{k} \to \mathfrak{d} \twoheadrightarrow \mathfrak{n}$

and an obvious map $\Delta \to \exp(\mathfrak{f}/[\mathfrak{f},\mathfrak{r}])$ giving a map $E_1 \to E(\mathfrak{n})$ by Malcev correspondence. The composition of these two maps are the identity on the generators of $\mathfrak{f}/[\mathfrak{f},\mathfrak{r}]$ and of Δ resp., hence on $\mathfrak{f}/[\mathfrak{f},\mathfrak{r}]$ and $\mathfrak{d}$. The composite isomorphism $H_2(\Gamma;\mathbb{Q}) \cong \exp \mathfrak{k} \cong H_2(\mathfrak{n})$ thus obtained is the map of the lemma, as follows immediately from the definition □

5.3.9 REMARK It has been proved by several authors [51, 41, cf. 37] that $H_i(N) \simeq H_i(\mathfrak{n})$ resp. $H_i(\Gamma;\mathbb{Q}) \simeq H_i(\mathfrak{n})$ for $i > 0$. We cannot just cite their result, because we need the isomorphism σ explicitly and it is not clear that their isomorphism is our σ for $i = 2$. Furthermore our proof follows easily from the Malcev correspondence theorem, which is an essential ingredient of all isomorphism proofs.

5.4 A TOPOLOGY ON THE HOPF EXTENSION

Let $\mathfrak{n}$ be a nilpotent Lie algebra over a p-adic field K. Let $\mathfrak{n}|k$ be the Lie algebra $\mathfrak{n}$ with scalars restricted to k, where k is a field with $\mathbb{Q} \subset k \subset K$. We define a topology on the Hopf extensions of $\mathfrak{n}|\mathbb{Q}$ and N resp. and prove that the exponential map is a homeomorphism. This will be needed to prove the desired continuity of the map $H_2(\mathfrak{n}|\mathbb{Q}) \to \ker \pi_U$ (see proof of 5.6.1).

5.4.1 Let $\mathfrak{n}$ be a finite-dimensional nilpotent Lie algebra over a p-adic field K. Let $R \rightarrowtail F \twoheadrightarrow N = \exp(\mathfrak{n})$ and $\mathfrak{r} \rightarrowtail \mathfrak{f} \twoheadrightarrow \mathfrak{n}|\mathbb{Q}$ be exact sequences of groups and Lie algebras over $\mathbb{Q}$ resp. with F and $\mathfrak{f}$ free. We have the exact sequences

E : $R/(F,R) \rightarrowtail F/(F,R) \twoheadrightarrow N$

whose derived sequence is the Hopf extension

E' : $H_2(N) \rightarrowtail F'/(F,R) \twoheadrightarrow N'$.

Similarly

$\mathfrak{E}$: $\mathfrak{r}/[\mathfrak{f},\mathfrak{r}] \rightarrowtail \mathfrak{f}/[\mathfrak{f},\mathfrak{r}] \twoheadrightarrow \mathfrak{n}$

and

$\mathfrak{E}'$: $H_2(\mathfrak{n}|\mathbb{Q}) \rightarrowtail \mathfrak{f}'/[\mathfrak{f},\mathfrak{r}] \twoheadrightarrow \mathfrak{n}'$.

There is a map of exact sequences $\varphi : E \to \exp \mathfrak{E}$ inducing the identity on N , whose derived map $\varphi' : E' \to \exp \mathfrak{E}'$ is independent of the map φ chosen. It is an isomorphism by theorem 5.3.7.

Let us denote $\mathfrak{d} = \mathfrak{f}/[\mathfrak{f},\mathfrak{r}]$, $D = \exp \mathfrak{d}$, $G = F/(F,R)$,

E : $R/(F,R) \rightarrowtail G \xrightarrow{\Pi} N$,

$\mathfrak{E}$: $\mathfrak{r}/[\mathfrak{f},\mathfrak{r}] \rightarrowtail \mathfrak{d} \xrightarrow{\pi} \mathfrak{n}$.

We shall define a topology on G' and $\mathfrak{d}'$ resp. such that $\varphi' : G' \to \exp \mathfrak{d}'$ becomes an isomorphism of topological groups. Recall the map $\xi : \mathfrak{n} \otimes_{\mathbb{Q}} \mathfrak{n} \to \mathfrak{d}'$ of 5.2.3 defined by

$$\xi(X \otimes Y) = [\pi^{-1}X, \pi^{-1}Y] .$$

The map ξ is well-defined since the kernel of π is central in $\mathfrak{d}$. Being of finite dimension over K , $\mathfrak{n}$ has a unique topology as a topological vector space over the local field K . The point is that $\mathfrak{d}'$ is a vector space only over $\mathbb{Q}$.

5.4.2 Lemma *There is a unique finest topology on $\mathfrak{d}'$ with the properties a) - c).*

a) $\mathfrak{n} \times \mathfrak{n} \to \mathfrak{n} \otimes_{\mathbb{Q}} \mathfrak{n} \xrightarrow{\xi} \mathfrak{d}'$ *is continuous.*

b) $\mathfrak{d}'$ *has a neighbourhood base of* 0 *consisting of additive subgroups of* $\mathfrak{d}'$ *.*

c) $\mathfrak{d}'$ *is a topological group with respect to addition.*

This topology has the following additional properties.

d) $\mathfrak{d}'$ *is a topological Lie algebra over* $\mathbb{Q}$ *, where* $\mathbb{Q}$ *carries the p-adic topology.*

e) The map $\pi : \mathfrak{d}' \to \mathfrak{n}$ *is continuous.*

Proof. For every subset V of $\mathfrak{n}$ let us denote by V^* the additive subgroup of $\mathfrak{d}'$ generated by the set $\{[X,Y]; \pi(X)\in V, \pi(Y)\in V\}$. Set

$$\mathfrak{V}^* = \{V^*;\ V \text{ a neighbourhood of } 0 \text{ in } \mathfrak{n}\}$$

and

$$\mathfrak{V}^*_{Lie} = \{V^*;\ V \text{ an open Lie subring of } \mathfrak{n}\} .$$

There are unique topologies $\mathfrak{T}$ and $\mathfrak{T}_{Lie}$ on $\mathfrak{d}'$ which satisfy c) and for which $\mathfrak{V}^*$ and $\mathfrak{V}^*_{Lie}$ resp. are a neighbourhood base of 0 . We have $\mathfrak{T} = \mathfrak{T}_{Lie}$, since the open Lie subrings of $\mathfrak{n}$ form a neighbourhood base of 0 in $\mathfrak{n}$, by 2.6.1. Clearly, $\mathfrak{T}$ satisfies a) - c) and is finer than every topology satisfying a) - c).

e) holds since $\mathfrak{V}^*_{Lie}$ is a neighbourhood base of 0 in $\mathfrak{d}'$.

d) The following claims are easily proved and imply d). The bilinear map $\mathbb{Q} \times \mathfrak{d}' \to \mathfrak{d}'$ given by multiplication with scalars is continuous at $(0,0)$ and continuous in each variable separately, hence continuous. Similarly for the Lie bracket $\mathfrak{d}' \times \mathfrak{d}' \to \mathfrak{d}'$ □

The exponential map $\exp : \mathfrak{d} \to D$ induces a bijection $\mathfrak{d}' \to D'$, by 2.5.7. So there is a unique topology on D' making $\exp : \mathfrak{d}' \to D'$ a homeomorphism. In the next lemma we shall give a different description of this topology. It will be useful since this description is given in the category of topological groups and does not make use of Lie algebras.

The Lie algebra homomorphism $\pi : \mathfrak{d} \to \mathfrak{n}$ induces a group homomorphism $\exp \pi : D \to N$. Again we have a map

$$\zeta : N \times N \longrightarrow D'$$

such that
$$\zeta(\exp \pi\, x, \exp \pi\, y) = (x,y)$$

for x,y in D , where the bracket on the right hand side denotes the (group) commutator of the elements x and y of D . The map ζ is well-defined, since $\ker \exp \pi = \exp \ker \pi$ is central in D .

The following lemma easily implies the main result of this section. Its proof relies heavily on the results of 2.5.

5.4.3 LEMMA *The topology of* D' *given by transport of structure via* $\exp : \mathfrak{d}' \to D'$ *is the finest topology with the following properties.*

f) ζ *is continuous.*

g) D' *has a neighbourhood base of* e *consisting of subgroups of* D' .

h) D' *is a topological group.*

Furthermore, this topology has the following property.

i) The following system $\mathfrak{V}_{gr}$ *of subgroups of* D' *is a neighbourhood base of* e *in* D' .

$$\mathfrak{V}_{gr} = \{((\exp \pi)^{-1}A)' \; ; \; A \textit{ an open subgroup of } N\} .$$

PROOF. Let $\mathfrak{T}$ be the topology on D' making $\exp : \mathfrak{d}' \to D'$ a homeomorphism. Let us first see that $\mathfrak{T}$ has the properties f) - h). The Campbell-Hausdorff formula and d) imply h).

f) For every Lie monomial L of degree $d \geq 2$ define a d-multilinear map $\xi_L : \mathfrak{n}\times\ldots\times\mathfrak{n} \to \mathfrak{d}'$, which generalizes the map $\mathfrak{n}\times\mathfrak{n} \to \mathfrak{n} \otimes_{\mathbb{Q}} \mathfrak{n} \xrightarrow{\xi} \mathfrak{d}'$ for the Lie monomial L with $L(X,Y) = [X,Y]$. Every such map ξ_L is continuous, as follows by induction on d : Start with a) and use the fact that $\mu : \mathfrak{n} \times \mathfrak{d}'$, $\mu(X,Y) = [\pi^{-1}X,Y]$ is continuous, since μ is the composition of the two continuous maps $\mathfrak{n} \times \mathfrak{d}' \xrightarrow{1\times\pi} \mathfrak{n} \times \mathfrak{n}$ and $\mathfrak{n} \times \mathfrak{n} \to \mathfrak{n} \otimes_{\mathbb{Q}} \mathfrak{n} \to \mathfrak{d}'$ of a) . The Campbell-Hausdorff formula shows that $\zeta(\exp\pi\,(\exp X),\ \exp\pi\,(\exp Y)) = \exp \zeta_1(X,Y)$ for X,Y in $\mathfrak{n}$, where ζ_1 is a rational linear combination of maps ξ_L with entries only X and Y. This proves that ζ is continuous.

g) is implied by 2.5.13 as follows. Let $V^* \in \mathfrak{V}^*$. Then for $c \in \mathbb{N}$ as in 2.5.13 for the step r of $\mathfrak{d}'$ let $U^* \in \mathfrak{V}^*_{Lie}$ be such that $c^{-1}\,U^* \subset V^*$. Note that U^* is a Lie subring of $\mathfrak{d}'$. So there is a Lie subring $\mathfrak{h}$ of $\mathfrak{d}'$ such that $\exp \mathfrak{h}$ is a subgroup of D' and such that

$$\mathfrak{h} \supset U^* \supset c\,\mathfrak{h} .$$

Therefore $\mathfrak{h}$ is a neighbourhood of 0 by the first inclusion and $\mathfrak{h} \subset c^{-1}\,U^* \subset V^*$ by the second inclusion.

Conversely, let $\mathfrak{T}'$ be a topology on D' with the properties f) - h). We have to show that $\mathfrak{T}' \subset \mathfrak{T}$. By h) it suffices to show that every $W \in \mathfrak{T}'$ containing e is in $\mathfrak{T}$. We may assume that W is a subgroup of D' , by g). We first show that for every group $W \in \mathfrak{T}'$ there is a neighbourhood V of 0 in $\mathfrak{n}$ such that $\exp \xi(X\otimes Y) \in W$ for X,Y in V. By 2.5.10 there is a number $c \in \mathbb{N}$ for the step r of nilpotency of $\mathfrak{d}$ such that

$$\exp[cA,cB] \in \langle \exp A\ ,\ \exp B\rangle' \quad (*)$$

for A, B in $\mathfrak{d}$. There is an open Lie subring V_1 of $\mathfrak{n}$ such that $\exp V_1$ is a subgroup of N and $\zeta(V_1 \times V_1) \subset W$, by f) and 2.6.1. Hence if X, Y are in $V = c\,V_1$, then $\exp \xi(X\otimes Y) = \exp[\pi^{-1}X, \pi^{-1}Y]$ is by $(*)$ contained in the commutator subgroup of the group $(\exp \pi)^{-1} \exp V_1$, which is contained in W.

We now finish the proof of the first part of the lemma by showing that for every group $W \in \mathfrak{T}'$ there is a neighbourhood V of 0 in $\mathfrak{n}$ such that $\exp V^* \subset W$, where V^* is as above the additive subgroup of $\mathfrak{d}'$ generated by the set $\xi(X\otimes Y)$, X, Y in V. There is a number $c \in \mathbb{N}$ such that for every subgroup W of $\mathfrak{d}'$ we have $c\cdot\langle\log W\rangle \subset \log W$, where $\langle\log W\rangle$ denotes the Lie subring of $\mathfrak{d}'$ generated by $\log W$. This is the special case of 2.5.13, where X is the logarithm of a group. By the last paragraph, for every group $W \in \mathfrak{T}'$ there is a neighbourhood V_1 of 0 in $\mathfrak{n}$ such that $\xi(X\otimes Y) \in \log W$ for X, Y in V_1. Hence for cX, cY in $V := cV_1$ we have $\xi(cX\otimes cY) = c^2\xi(X\otimes Y) \in c^2 \log W$, so the additive group V^* is contained in $c^2\langle\log W\rangle \subset c\langle\log W\rangle \subset \log W$.

We now prove the additional property i) of the topology $\mathfrak{T}$ on D'.
i) The system $\mathfrak{V}_{gr}$ of subgroups of D' is mapped into itself by inner automorphisms of D. So there is a unique topology $\mathfrak{T}'$ on D' for which h) and i) hold. Let $W \in \mathfrak{T}$ be a subgroup of D'. Then there is an open subgroup A of N such that $\zeta(A\times A) \subset W$, by f) and 2.6.1. Hence the commutator subgroup of $(\exp \pi)^{-1} A$ is contained in W. So $\mathfrak{T}' \supset \mathfrak{T}$. Conversely, $\mathfrak{T}'$ has property g) and $\zeta : N \times N \to D'$ is continuous at (e,e) for $\mathfrak{T}'$, obviously, hence ζ is continuous as follows easily from the "bilinearity formulae" 2.2.3, 2) and 2'). So $\mathfrak{T}'$ is coarser than $\mathfrak{T}$ □

Recall the group $G = F/(F,R)$ and the group homomorphism $\Pi : G \to N$

of 5.4.1. Define a system $\mathfrak{D}_{gr}$ of subgroups of G' by $\mathfrak{D}_{gr} = \{\Pi^{-1}(A))'$; A an open subgroup of N} . The system $\mathfrak{D}_{gr}$ is invariant under inner automorphisms of G . Hence there is a unique topology on G' such that G' is a topological group and $\mathfrak{D}_{gr}$ is a neighbourhood base of e .

5.4.4 PROPOSITION *The map* $\varphi' : G' \to D'$ *of 5.4.1 is an isomorphism of topological groups.*

PROOF. Theorem 5.3.7 shows that φ' is an isomorphism of groups. It is a homeomorphism, since the topology of D' is characterized by 5.4.3 h) and i) □

5.5 $H_2(\mathfrak{n}|\mathbb{Q}_p)^0 \cong H_2(\mathfrak{n}|K)^0$

In this section we prove a lemma to be used in the proof of the main theorem, namely the formula of the title, proposition 5.5.4. It shows that restricting scalars of $\mathfrak{n}$ to $\mathbb{Q}_p$ does not change the part of the second homology we are interested in, if the necessary condition of chapter III is satisfied.

In order to prove the proposition we have to discuss the action of Q on $\Lambda(\mathfrak{n})$. This gives another lemma, corollary 5.5.3, to be used in the proof of the main theorem. The notations of this section will be used in the next section in the proof of the main theorem.

5.5.1 Again let $\mathfrak{n}$ be a finite dimensional nilpotent Lie algebra over a p-adic field K . Let the finitely generated free abelian group Q act on $\mathfrak{n}$ by automorphisms. Suppose (cf. 3.1.5)

$$\mathfrak{n} = \oplus\, \mathfrak{n}^{\alpha} \ , \ \alpha \in \mathrm{Hom}(Q,\mathbb{R}) \ .$$

For every field $k \subset K$ let $\otimes_k^s \mathfrak{n}$ be the s-fold tensor product of $\mathfrak{n}|k$ over the field k . For every $\beta \in \mathrm{Hom}(Q,\mathbb{R})$ define the subspace $(\otimes_k^s \mathfrak{n})^{\beta}$ of $\otimes_k^s \mathfrak{n}$ as the sum over all subspaces $\mathfrak{n}^{\alpha_1} \otimes_k \ldots \otimes_k \mathfrak{n}^{\alpha_s}$ with $\alpha_1 + \ldots + \alpha_s = \beta$.

We have

$$\otimes_k^s \mathfrak{n} = \oplus (\otimes_k^s \mathfrak{n})^\beta$$

and an induced decomposition

$$\Lambda_k^s \mathfrak{n} = \oplus (\Lambda_k^s \mathfrak{n})^\beta \quad , \quad \beta \in \mathrm{Hom}(Q, \mathbb{R}) \quad ,$$

since a permutation of the tensor factors maps every summand $(\otimes_k^s \mathfrak{n})^\beta$ into itself. The differential $d_i : \Lambda_k^i \mathfrak{n} \to \Lambda_k^{i-1} \mathfrak{n}$ of the Koszul complex (see 5.2.1) respects the decompositions, hence

$$\ker d_i =: Z_i(k) = \oplus Z_i(k)^\beta \quad ,$$

where

$$Z_i(k)^\beta := Z_i(k) \cap (\Lambda_k^i \mathfrak{n})^\beta \quad ,$$

and

$$\mathrm{im}\, d_{i+1} =: B_i(k) = \oplus B_i(k)^\beta \quad ,$$

where

$$B_i(k)^\beta := B_i(k) \cap (\Lambda_k^i \mathfrak{n})^\beta \quad ,$$

and hence

$$\frac{Z_i(k)}{B_i(k)} = H_i(\mathfrak{n}|k) = \oplus H_i(\mathfrak{n}|k)^\beta \quad ,$$

where

$$H_i(\mathfrak{n}|k)^\beta = \frac{Z_i(k)^\beta}{B_i(k)^\beta} \quad .$$

For $k = K$ these definitions coincide with our earlier notations, i.e. $V^\beta = \oplus V^\alpha$, $\nu_*(\alpha) = \beta$, $\alpha \in \mathrm{Hom}(Q, K^*)$, and $V^\alpha = \{x \in V; tx = \alpha(t)x$ for every $t \in Q\}$, but for $k \subsetneqq K$ the expression $\alpha(t)x$ makes no sense, since we have vector spaces only over k . Now suppose $k \supset \mathbb{Q}_p$. For any norm on the finite dimensional topological vector space $\otimes_k^s \mathfrak{n}$ over k , let $\|t\|_{s,\gamma}$ be the norm of the map of $(\otimes_k^s \mathfrak{n})^\gamma$ to itself corresponding to $t \in Q$. We denote by $e > 1$ the basis of the logarithm in the definition of ν (see 3.1.4).

5.5.2 LEMMA *Given* s , *there is a constant* c *such that*

$$\|t\|_{s,\gamma} \le c \cdot e^{\gamma(t)}$$

for every $\gamma \in \mathrm{Hom}(Q, \mathbb{R})$ *and* $t \in Q$.

PROOF. It suffices to prove our claim for one norm, since any two norms on a finite dimensional vector space over k are equivalent. If $V \subset \mathfrak{n}^\alpha$

is a one-dimensional normed vector subspace over K , then t has norm $e^{\alpha(t)}$ on V . Our claim is now implied by the following fact. Given two normed vector spaces V , W and linear endomorphisms $A \in \text{End}(V)$, $B \in \text{End}(W)$. Then, for the norm

$$\|z\| = \inf \{\Sigma\|v_i\| \cdot \|w_i\| \ , \ z = \Sigma v_i \otimes w_i \ , \ v_i \in V \ , \ w_i \in W\}$$

on $V \otimes_K W$, we have $\|A \otimes B\| \leq \|A\| \cdot \|B\|$ $\square$

5.5.3 COROLLARY
If $\gamma \neq 0$ *, then zero is in the closure of the Q-orbit of every element of* $(\otimes_k^s \mathfrak{n})^\gamma$ *and hence of every element of* $(\wedge_k^s \mathfrak{n})^\gamma$ $\square$

5.5.4 PROPOSITION
Suppose $K \supset k \supset \mathbb{Q}_p$ *. If* $R \cap -R = \emptyset$ *, the map*
$$H_2(\mathfrak{n}|k)^o \to H_2(\mathfrak{n})^o$$
induced by the inclusion of coefficients is an isomorphism.

The field of coefficients at the target of the arrow is K .

5.5.5
Note that the claim of this proposition is not true if $R \cap -R \neq \emptyset$, e.g. if $R = \{0\}$ and $\mathfrak{n}$ is abelian, then $H_2(\mathfrak{n}|k)^o = \mathfrak{n} \wedge_k \mathfrak{n}$, but $H_2(\mathfrak{n})^o = \mathfrak{n} \wedge_K \mathfrak{n}$. It is also not true for $k = \mathbb{Q}$, even if any two elements of R are positively independent, see 5.7.4.

PROOF OF 5.5.4.
We shall show that the composed linear map τ
$$\mathfrak{n} \otimes_k \mathfrak{n} \to \wedge_k^2 \mathfrak{n} \to \wedge_k^2 \mathfrak{n}/B_2(k)$$
vanishes on $\lambda z \otimes y - z \otimes \lambda y$ for $\lambda \in K$ and z or y in $\mathfrak{n}'$.

I show first that this claim implies the proposition. If the claim holds, then τ induces a k-linear map $\bar{\tau} : (\mathfrak{n} \otimes_K \mathfrak{n})^o = \Sigma \mathfrak{n}^\alpha \otimes_K \mathfrak{n}^{-\alpha} \to \wedge_k^2 \mathfrak{n}/(B_2(k)$, since for any $\alpha \in \text{Hom}(Q, \mathbb{R})$ one of the two spaces $\mathfrak{n}^\alpha$, $\mathfrak{n}^{-\alpha}$ is contained in $\mathfrak{n}'$, by our hypotheses $R \cap -R = \emptyset$. The map $\bar{\tau}$ vanishes on the subspace generated by $X \otimes Y - Y \otimes X$, $X \in \mathfrak{n}^\alpha$, $Y \in \mathfrak{n}^{-\alpha}$, which is the intersection of the kernel of $\mathfrak{n} \otimes_K \mathfrak{n} \to \mathfrak{n} \wedge_K \mathfrak{n}$, $X \otimes Y \to X \wedge Y$, with $(\mathfrak{n} \otimes_K \mathfrak{n})^o$. So $\bar{\tau}$ induces a k-linear map $\bar{\bar{\tau}} : (\mathfrak{n} \wedge_K \mathfrak{n})^o \to \wedge_k^2 \mathfrak{n}/B_2(k)$, which vanishes on $B_2(K)^o$, hence induces a k-linear map

$(\mathfrak{n} \wedge_K \mathfrak{n})^O/B_2(K)^O \to (\Lambda_k^2\mathfrak{n})^O/B_2(k)^O$, which is a left inverse of the map $H_2(\mathfrak{n}|k)^O \to H_2(\mathfrak{n})^O$ induced by restriction of scalars. Surjectivity of $H_2(\mathfrak{n}|k)^O \to H_2(\mathfrak{n})^O$ follows from the fact, that $Z_2(k) \to Z_2(K)$ is surjective and preserves the decomposition $\oplus\ldots^\gamma$ according to the type $\gamma \in \mathrm{Hom}(Q,\mathbb{R})$. This proves the proposition if the claim holds.

We now prove the claim. Define

$$A = \Lambda_k^2\mathfrak{n}/B_2(k) .$$

It suffices to prove that

$$\tau(\lambda X_1 \otimes [X_2,X_3]) = \tau(X_1 \otimes \lambda[X_2,X_3])$$

holds for X_1,X_2,X_3 in $\mathfrak{n}$, $\lambda \in K$, since τ is linear and alternating. Define three mappings

$$\varphi_i : K \otimes_k K \to A , \quad i = 1,2,3$$

by

$$\begin{aligned}\varphi_i(\lambda \otimes \mu) &= \tau(\lambda X_i \otimes \mu[X_{i+1},X_{i+2}]) \\ &= \lambda X_i \wedge \mu[X_{i+1},X_{i+2}] \bmod B_2(k) ,\end{aligned}$$

taking indices modulo 3. We have

$J(\lambda,\mu,\nu)$: $$\varphi_1(\lambda\otimes\mu\nu) + \varphi_2(\mu\otimes\nu\lambda) + \varphi_3(\nu\otimes\lambda\mu) = 0$$

by definition of $B_2(k)$ and since the Lie bracket in $\mathfrak{n}$ is K-bilinear. We have to show that the maps φ_i are K-bilinear. For this purpose define

$$\Psi_i : K \otimes_k K \otimes_k K \to A$$

by

$$\Psi_i(\lambda \otimes \mu \otimes \alpha) := \varphi_i(\alpha\lambda \otimes \mu) - \varphi_i(\lambda \otimes \alpha\mu) .$$

We have to show that $\Psi_i = 0$. The equation $J(\lambda,\alpha\mu,\nu)$ minus $J(\lambda,\mu,\alpha\nu)$ yields

$$\Psi_2(\mu \otimes \nu\lambda \otimes \alpha) = \Psi_3(\nu \otimes \lambda\mu \otimes \alpha) \tag{1}$$

for μ,ν,λ and α in K. This yields for $\lambda = 1$

$$\Psi_2(\mu \otimes \nu \otimes \alpha) = \Psi_3(\nu \otimes \mu \otimes \alpha) \tag{2}$$

Combine (2) and (1) to obtain

$$\Psi_2(\lambda\mu \otimes \nu \otimes \alpha) = \Psi_3(\nu \otimes \lambda\mu \otimes \alpha) = \Psi_2(\mu \otimes \nu\lambda \otimes \alpha) .$$

So Ψ_2 is, with respect to the first two variables, K-bilinear, hence symmetric. So (2) implies $\Psi_2 = \Psi_3$. Similarly

$$\Psi_1 = \Psi_2 = \Psi_3 \tag{3}$$

So there is a well defined k-linear map

$$\rho : K \otimes_k K \to A ,$$

such that $\rho(\lambda\mu \otimes \nu) = \Psi_i(\lambda \otimes \mu \otimes \nu)$. (4)

Compute $\rho(\lambda \otimes \mu\nu) = \Psi_1(\lambda \otimes 1 \otimes \mu\nu)$

$$= \varphi_1(\lambda\mu\nu \otimes 1) \qquad\qquad\qquad\qquad - \varphi_1(\lambda \otimes \mu\nu)$$

$$= \varphi_1(\lambda\mu\nu \otimes 1) - \varphi_1(\lambda\mu \otimes \nu) + \varphi_1(\lambda\mu \otimes \nu) - \varphi_1(\lambda \otimes \mu\nu)$$

$$= \Psi_1(\lambda\mu \otimes 1 \otimes \nu) + \Psi_1(\lambda \otimes \nu \otimes \mu) = \rho(\lambda\mu \otimes \nu) + \rho(\lambda\nu \otimes \mu) .$$

So

$$\rho(\lambda \otimes \mu\nu) = \rho(\lambda\mu \otimes \nu) + \rho(\lambda\nu \otimes \mu) . \tag{5}$$

We have to show that $\Psi_i = 0$ or equivalently that $\rho = 0$. For $\alpha \in k$ we have $\rho(\lambda \otimes \alpha) = 0$ by (5) and the k-bilinearity of ρ . Equation (5) yields by induction on n

$$\rho(\lambda \otimes \mu^n) = \rho(\lambda \cdot n \cdot \mu^{n-1} \otimes \mu) .$$

Therefore, if f is a polynomial with coefficients in k , we have for λ,μ in K

$$\rho(\lambda \otimes f(\mu)) = \rho(\lambda f'(\mu) \otimes \mu) . \tag{6}$$

Let now μ be an element of K . Since K is separable algebraic over k , there is a polynomial f with coefficients in k , such that $f(\mu) = 0$ and $f'(\mu) \neq 0$. Equation (6) implies now $\rho(\lambda \otimes \mu) = 0$ for every $\lambda \in K$. So $\rho = 0$, q.e.d. □

5.6 The main theorem

In this section we prove the main theorem 5.6.1. Necessity has been proved earlier. Sufficiency follows from the main result of chapter IV and the comparsion results for the second homology of this chapter. Examples and counterexamples will be given in the next section 5.7. The main theorem will be applied for algebraic groups over local fields and S-arithmetic groups in the next two chapters VI and VII.

For the sake of reference we recall the hypotheses and notations.

Let K be a p-adic field, let $\mathfrak{n}$ be a nilpotent Lie algebra over K of finite dimension. Let $N = \exp \mathfrak{n}$ be the corresponding Lie group. Let Q be a finitely generated free abelian group acting on $\mathfrak{n}$ by automorphisms. For every $\lambda \in \mathrm{Hom}(Q,K^*)$ define

$$\mathfrak{n}^\lambda = \{X \in \mathfrak{n} \; ; \; tX = \lambda(t)\cdot X \quad \text{for every} \quad t \in Q\}$$

and similarly for $\mathfrak{n}^{ab}$. We suppose

$$\mathfrak{n} = \oplus \, \mathfrak{n}^\lambda \; , \; \lambda \in \mathrm{Hom}(Q,K^*) \; .$$

We set $W(\mathfrak{n}^{ab}) = \{\lambda \in \mathrm{Hom}(Q,K^*) \; ; \; (\mathfrak{n}^{ab})^\lambda \neq 0\}$.

Let $\qquad \nu_* : \mathrm{Hom}(Q,K^*) \to \mathrm{Hom}(Q,\mathbb{R})$

be the map $\qquad \nu_*(\lambda) = -\log \circ \mathrm{mod}_K \circ \lambda$,

where mod_K is the natural valuation of K (see 1.2). Set

$$R = \nu_*(W(\mathfrak{n}^{ab})) \; .$$

We denote by $H_2(\mathfrak{n})^o$ the subspace of $H_2(\mathfrak{n})$ of vectors having relatively compact orbit with respect to the action of Q on $H_2(\mathfrak{n}) := H_2(\mathfrak{n};K)$ induced by the action of Q on $\mathfrak{n}$. We have (see 5.2.4)

$$H_2(\mathfrak{n})^o = \bigoplus_{\nu_*(\lambda)=0} H_2(\mathfrak{n})^\lambda$$

5.6.1 THEOREM $Q \ltimes N$ *has a compact presentation iff the following two conditions hold.*

1) Any two elements of R *are positively independent, i.e. for any two vectors of* R *the zero vector is not in their convex hull.*

2) $\qquad H_2(\mathfrak{n})^o = 0$.

Note that the two elements of R in 1) need not be different. So 1) implies - and is for $\# R = 1$ equivalent to - $0 \notin R$, a necessary and sufficient condition for $Q \ltimes N$ to have a compact set of generators, see 3.2.2.

For abelian N the weights on $H_2(\mathfrak{n}) = \mathfrak{n} \wedge \mathfrak{n}$ are the sums $\alpha + \beta$ of elements α, β in $R = \nu_* W(\mathfrak{n})$, hence

5.6.2 COROLLARY *If* N *is furthermore abelian, then* $Q \ltimes N$ *has a compact presentation iff 1) holds.*

For N step 2 nilpotent, condition 1) is not sufficient for compact presentability of $Q \ltimes N$, see 3.3.3, 5.7.1.

Since $H_2(\mathfrak{n})$ is a subquotient of the Q-module $\mathfrak{n} \wedge \mathfrak{n}$ we have for the sets of weights $W(H_2(\mathfrak{n})) \subset W(\mathfrak{n}\wedge\mathfrak{n}) \subset \{\alpha+\beta \mid \alpha,\beta \in W(\mathfrak{n})\}$, hence

5.6.3 REMARK *Condition 2) holds if zero is not the sum of any two elements of* $\nu_* W(\mathfrak{n})$.

PROOF OF 5.6.1 Necessity of the two conditions has been shown in 3.2.3 and 5.2.5 resp. In chapter IV it was shown that under condition 1) either $Q \ltimes N$ has a compact presentation or the central extension $\pi_U : M_U \to N$ of 4.6 has a non-trivial kernel. So it remains to show that in the latter case we have $H_2(\mathfrak{n})^{o} \neq 0$. We shall freely use the notations of 5.4 and 5.5. Look at the map $\omega : \mathfrak{n}\otimes_{\mathbb{Q}}\mathfrak{n} \to M_U$ defined by composing the following maps

$$\mathfrak{n} \otimes_{\mathbb{Q}} \mathfrak{n} \xrightarrow{\xi} \mathfrak{d}' \xrightarrow[\sim]{\exp} D' \xleftarrow[\sim]{\varphi'} G' \xrightarrow{\eta} M_U' .$$

Here $\mathfrak{d}'$, D', G', ξ and φ' are as in 5.4, η is the map of the Hopf extension for M_U corresponding to the central extension $\pi_U : M_U \to N$, see 5.1.2. The Jacobi identity implies that

$$\xi([X,Y] \otimes Z + [Y,Z] \otimes X + [Z,X] \otimes Y) = 0 .$$

We shall show that ω induces a surjective map $H_2(\mathfrak{n})^{o} \to \ker \pi_U$, thus proving the theorem. Note that ω is not a homomorphism, in general, since exp is not.

We first show that ω induces a map $\mathfrak{n} \wedge_{\mathbb{Q}_p} \mathfrak{n} \to M_U'$. There is a unique topology on $\mathfrak{n} \otimes_{\mathbb{Q}} \mathfrak{n}$ such that $\mathfrak{n} \otimes_{\mathbb{Q}} \mathfrak{n}$ is a topological group with respect to addition and the set of images of the maps $V \otimes_{\mathbb{Z}} V \to \mathfrak{n} \otimes_{\mathbb{Q}} \mathfrak{n}$ is a neighbourhood base of zero, where V runs through all open Lie subrings of $\mathfrak{n}$. With respect to this topology ω is continuous, since all its constituents are continuous: ξ by definition of the topologies on $\mathfrak{n} \otimes_{\mathbb{Q}} \mathfrak{n}$ and $\mathfrak{d}'$ resp., see 5.4, φ' by proposition 5.4.4 and η, because for any open subgroup W of U we have $\eta(\Pi^{-1}W))' \subset (W \cdot \ker \pi_U)' \subset W'$, since $\ker \pi_U$ is central in M_U. The space $\mathfrak{n} \otimes_{\mathbb{Q}} \mathfrak{n}$ can be turned into a vector space over K in two ways, by left multiplication $\alpha(X \otimes Y) := (\alpha X) \otimes Y$ and by right multiplication $(X \otimes Y)\alpha := X \otimes (\alpha Y)$ for $\alpha \in K$, X,Y in $\mathfrak{n}$. The structure maps $K \times (\mathfrak{n} \otimes_{\mathbb{Q}} \mathfrak{n}) \to \mathfrak{n} \otimes_{\mathbb{Q}} \mathfrak{n}$ are actually continuous, so $\mathfrak{n} \otimes_{\mathbb{Q}} \mathfrak{n}$ thus becomes a topological vector space over K in two ways. The maps $K \to M'_U$, $\alpha \to \omega(\alpha Z)$ and $\alpha \to \omega(Z\alpha)$ are continuous for every $Z \in \mathfrak{n} \otimes_{\mathbb{Q}} \mathfrak{n}$. It follows that ω is constant on the cosets of the subspace $\{\Sigma\alpha_i X_i \otimes Y_i - X_i \otimes \alpha_i Y_i ; \alpha_i \in \mathbb{Q}_p , X_i , Y_i \text{ in } \mathfrak{n}\}$, since $\mathbb{Q}_p$ is the closure of $\mathbb{Q}$ in K, hence ω induces a map

$$\bar{\omega} : \mathfrak{n} \wedge_{\mathbb{Q}_p} \mathfrak{n} \to M'_U .$$

Note that $\bar{\omega}$ is continuous with respect to the unique topology of $\mathfrak{n} \wedge_{\mathbb{Q}_p} \mathfrak{n}$ making it a topological vector space over $\mathbb{Q}_p$.

Recall that ω and $\bar{\omega}$ are not homomorphisms, since exp is not. Let k be a field with $\mathbb{Q} \subset k \subset K$. Recall that we have defined $Z_2(k)$ as the kernel of $d_2 : \mathfrak{n} \wedge_k \mathfrak{n} \to \mathfrak{n}$ defined by $d_2(X \wedge Y) = [X,Y]$. Note that $Z_2(\mathbb{Q})$ is the inverse image under ξ of the kernel of $\pi : \mathfrak{d}' \to \mathfrak{n}'$. Since $\ker \pi$ is central in $\mathfrak{d}'$, exp induces a group isomorphism $\ker \pi \to \ker(\exp \pi : D' \to N')$, hence $\omega | Z_2(\mathbb{Q}) : Z_2(\mathbb{Q}) \to \ker \pi_U$ is a homomorphism of abelian groups. It is surjective, since all the maps occurring in the definition of ω are surjective and $\ker \pi_U \subset M'_U$, by 4.6.1 b). So we obtain a surjective homomorphism $\bar{\omega} : Z_2(\mathbb{Q}_p) \to \ker \pi_U$, since for any two fields $k_1 \subset k_2$ with $\mathbb{Q} \subset k_1 \subset k_2 \subset K$ the natural surjection $\mathfrak{n} \wedge_{k_1} \mathfrak{n} \to \mathfrak{n} \wedge_{k_2} \mathfrak{n}$ induces a surjection $Z_2(k_1) \to Z_2(k_2)$.

The group Q acts on every one of the groups and Lie algebras occurring in the maps which compose to give ω, and all these maps are Q-equivariant, hence so is ω and hence $\bar{\omega}$. Let $\gamma \in \mathrm{Hom}(Q,\mathbb{R})$ be non-zero. I show that $\bar{\omega}$ vanishes on $Z_2(\mathbb{Q}_p)^\gamma$. For every $z \in Z_2(\mathbb{Q}_p)^\gamma$ the closure of the Q-orbit contains zero, by 5.5.3, hence the closure of the Q-orbit of the image $\bar{\omega}(z)$ in the discrete group $\ker \pi_U$ contains zero, hence $\bar{\omega}(z) = 0$. So $\bar{\omega}$ restricts to a surjection $\bar{\omega} | Z_2(\mathbb{Q}_p)^0 \to \ker \pi_U$.

The map ω vanishes on $B_2(\mathbb{Q})$, since ξ does by the Jacobi identity, hence $\bar{\omega}$ vanishes on $B_2(\mathbb{Q}_p)$, the image of the map $B_2(\mathbb{Q}) \to B_2(\mathbb{Q}_p)$ induced by inclusion of the fields of coefficients. We thus obtain a surjection $\bar{\bar{\omega}} : H_2(\mathfrak{n}|\mathbb{Q}_p)^O \to \ker \pi_U$. This together with proposition 5.5.4, that $H_2(\mathfrak{n}|\mathbb{Q}_p)^O \cong H_2(\mathfrak{n})^O$, proves that ω induces a surjection $H_2(\mathfrak{n})^O \to \ker \pi_U$, q.e.d □

5.7 EXAMPLES

In this section we give two examples. We also give a third description of $H_2(\mathfrak{n}|k)^O$, besides the two in 5.2. As a corollary we obtain a compactly presentable extension $Q \ltimes \exp \mathfrak{m} \to Q \ltimes \exp \mathfrak{n}$, if $\mathfrak{n}$ satisfies condition 1), see 5.7.3.

First we show in 5.7.1 that the group of 3.3.1 actually has a compact presentation. Using the main theorem this is a trivial verification. In 5.7.4 we give an example where $Q \ltimes \exp \mathfrak{n}$ has a compact presentation, but $H_2(\mathfrak{n}|\mathbb{Q})^O \to H_2(\mathfrak{n})^O$ is not an isomorphism, hence $\varinjlim N^C \to N$ is not an isomorphism. So in this section we keep several promises made earlier (3.3.1, 3.3.2, introduction of 4.5, 5.5.5).

5.7.1 EXAMPLE Let Q, $\mathfrak{n}$ be as in 3.3.1. We show that $Q \ltimes \exp \mathfrak{n}$ has a compact presentation, as claimed in 3.3. Any two elements of R are positively independent, as shown in 3.3.1. Recall that

$$\mathfrak{n} = \oplus\, \mathfrak{n}^{\alpha} \;,\; \alpha \in \mathrm{Hom}(Q,\mathbb{R})$$

with $\nu_*W(\mathfrak{n}) = \{\delta_i - \delta_j \,,\, 1 \le i < j \le 4\}$ and $\delta_1 = \delta_4 = 0$.

In the following picture we have denoted at the point with coordinates (a,b) a basis vector of $\mathfrak{n}^{a\delta_2+b\delta_3}$.

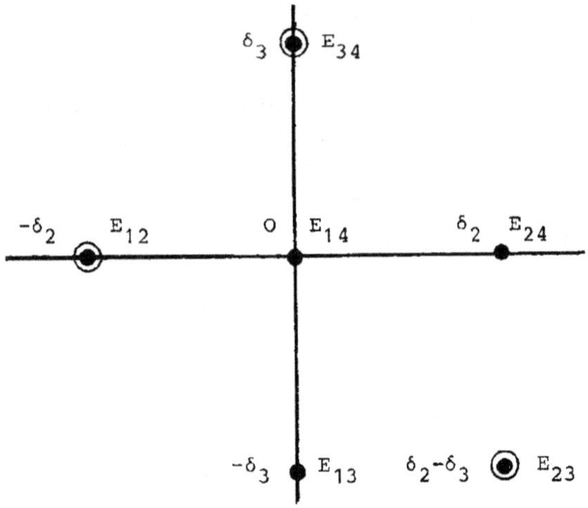

Now $(\mathfrak{n}\wedge\mathfrak{n})^O$ is spanned by $E_{12}\wedge E_{24}$, $E_{13}\wedge E_{34}$, hence $(\ker d_2)^O$ (see 5.2) is spanned by

$$E_{12}\wedge E_{24} - E_{13}\wedge E_{34} = d_3\,(E_{12}\wedge E_{23}\wedge E_{34}) ,$$

hence $H_2(\mathfrak{n})^O = O$ □

We next want to give an example showing that $H_2(\mathfrak{n}|\mathbb{Q})^O \to H_2(\mathfrak{n})^O$ and $\varinjlim N^C \to N$ need not be isomorphisms, even if $Q \ltimes N$ has a compact presentation. In order to do this we give a third description of $H_2(\mathfrak{n})^O$, after the one by the Koszul-complex and Hopf's description, both in 5.2.

Recall from 4.7 that if any two elements of R are positively independent then $\varinjlim \mathfrak{n}^C|k \to \mathfrak{n}$ is a central extension of $\mathfrak{n}$. We thus have a natural map $H_2(\mathfrak{n}|k) \to \ker(\varinjlim \mathfrak{n}^C|k \to \mathfrak{n})$, by 5.2.2. Recall the notations of 5.2.4.

5.7.2 PROPOSITION *If any two elements of* R *are positively independent, then* $H_2(\mathfrak{n}|k)^O \to \ker(\varinjlim \mathfrak{n}^C|k \to \mathfrak{n})$ *is an isomorphism for* $\mathbb{Q} \subset k \subset K$.

PROOF. Let V be a $K[Q]$-submodule of $\mathfrak{n}$, such that the natural map $\mathfrak{n} \to \mathfrak{n}^{ab}$ induces an isomorphism $V \cong \mathfrak{n}^{ab}$ of $K[Q]$-modules. Let f be a Lie algebra over k together with a k-linear map $V \to \mathfrak{f}$ with the following property. Every k-linear map $V \to \mathfrak{g}$ of V into a Lie algebra over k fits into a commutative diagram

$$\begin{array}{ccc} & & \mathfrak{f} \\ & \nearrow & \downarrow \\ V & & \\ & \searrow & \downarrow \\ & & \mathfrak{g} \end{array}$$

with a unique homomorphism $\mathfrak{f} \to \mathfrak{g}$ of Lie algebras over k. So $\mathfrak{f}$ may be regarded as a free Lie algebra over k with basis a basis of the k-vector space V. The k-linear map $V \to \mathfrak{n}$ induces a homomorphism $\vartheta : \mathfrak{f} \to \mathfrak{n}$ of Lie algebras over k. Let $\mathfrak{r}$ be the kernel of ϑ. We have an exact sequence

$$\mathfrak{r}/[\mathfrak{r},\mathfrak{f}] \rightarrowtail \mathfrak{f}/[\mathfrak{r},\mathfrak{f}] \twoheadrightarrow \mathfrak{n}$$

with $\mathfrak{r}/[\mathfrak{r},\mathfrak{f}] \simeq H_2(\mathfrak{n}|k)$, since $\mathfrak{f}^{ab} \xrightarrow{\sim} \mathfrak{n}^{ab}$, hence $\mathfrak{r} \subset \mathfrak{f}'$.

The representation of Q on V induces a representation of Q on $\mathfrak{f}$, $\mathfrak{r}$, $[\mathfrak{r},\mathfrak{f}]$ and $H_2(\mathfrak{n}|k)$ and the Hopf isomorphism $\mathfrak{r}/[\mathfrak{r},\mathfrak{f}] \simeq H_2(\mathfrak{n}|k)$ is Q-equivariant, by naturality. We have $V = \oplus V^{\beta}$, $\beta \in R$. There is a unique gradation $\mathfrak{f} = \oplus \mathfrak{f}^{\gamma}$, $\gamma \in \mathrm{Hom}(Q,\mathbb{R})$, of the Lie algebra $\mathfrak{f}$ such that $\mathfrak{f}^{\gamma} \cap V = V^{\gamma}$. We have induced gradations of $\mathfrak{r}$, $[\mathfrak{f},\mathfrak{r}]$ and $H_2(\mathfrak{n}|k)$, since $\mathfrak{n} = \oplus \mathfrak{n}^{\gamma}$ and ϑ respects the gradations. Then $\mathfrak{s} := \oplus H_2(\mathfrak{n}|k)^{\gamma}$, $\gamma \neq 0$, is a central ideal of $\mathfrak{f}/[\mathfrak{f},\mathfrak{r}]$. We define a central extension of $\mathfrak{n}$

$$\mathcal{E}_0 : \qquad H_2(\mathfrak{n}|k)^0 \rightarrowtail \mathfrak{h} \xrightarrow{\pi} \mathfrak{n}$$

by taking $H_2(\mathfrak{n}|k) \rightarrowtail \mathfrak{f}/[\mathfrak{r},\mathfrak{f}] \twoheadrightarrow \mathfrak{n}$ modulo $\mathfrak{s}$.

Let $\mathfrak{m}$ be the colimit $\varinjlim \mathfrak{n}^C|k$ of 4.7 and let $\Phi : \cup \mathfrak{n}^C \to \mathfrak{m}$ be the corresponding map. Let

$$\mathcal{E}_{\lim} : \qquad \mathfrak{k} \rightarrowtail \mathfrak{m} \twoheadrightarrow \mathfrak{n}$$

be the corresponding extension of $\mathfrak{n}$, which is central by 4.7. There is a unique k-linear map $\xi : V \to \mathfrak{m}$ such that $\xi|(V \cap \mathfrak{n}^C) = \Phi|(V \cap \mathfrak{n}^C)$ for every $C \in \mathcal{C}$, e.g. ξ may be defined as the linear map such that $\xi|V^{\beta} = \Phi|V^{\beta}$ for every $\beta \in R$. The map ξ induces a Q-equivariant Lie algebra homomorphism $\mathfrak{f} \to \mathfrak{m}$, also called ξ, which vanishes on $[\mathfrak{r},\mathfrak{f}]$

and hence induces a map $\bar{\xi} : \mathfrak{f}/[\mathfrak{r},\mathfrak{f}] \to \mathfrak{m}$. The map ξ , and hence $\bar{\xi}$, respects the gradations of $\mathfrak{f}$, $\mathfrak{f}/[\mathfrak{r},\mathfrak{f}]$, and $\mathfrak{m} = \oplus\, \mathfrak{m}^{\gamma}$, $\gamma \in \mathrm{Hom}(Q,\mathbb{R})$, by 4.7.2. The map $\bar{\xi}$ vanishes on $\mathfrak{s}$, since $\mathfrak{k} \subset \mathfrak{m}^{0}$, hence induces a map $\bar{\bar{\xi}} : E_0 \to E_{\lim}$ of exact sequences. The map $\bar{\bar{\xi}} : \mathfrak{h} \to \mathfrak{m}$ is surjective, since it induces the composite isomorphism $\mathfrak{h}^{ab} \cong \mathfrak{n}^{ab} \cong \mathfrak{m}^{ab}$, see 4.7.

We now show that there is a left inverse map of $\bar{\bar{\xi}}$, thus proving the proposition. The inverse image $\pi^{-1}(\mathfrak{n}^C)$ of $\mathfrak{n}^C$ in $\mathfrak{h}$ is a graded Lie subalgebra $\mathfrak{t} = \bigoplus_{\gamma\in C} \mathfrak{t}^{\gamma} \oplus H_2(\mathfrak{n}|k)^{0}$, since $0 \notin C$. Then $\mathfrak{t}^{+} = \bigoplus_{\gamma\in C} \mathfrak{t}^{\gamma}$ is a subalgebra of $\mathfrak{t}$, since C is a semigroup, and $\pi : \mathfrak{t}^{+} \to \mathfrak{n}^C$ is an isomorphism of Lie algebras over k , whose inverse we denote by η^C . Define a map $\eta : U\mathfrak{n}^C \to \mathfrak{h}$ by letting $\eta(x)$ be the element of $\pi^{-1}(x)$ whose component of degree 0 is zero. Then $\eta|\mathfrak{n}^C = \eta^C$ is a Lie algebra homomorphism over k and hence induces a Lie algebra homomorphism $\mathfrak{m} \to \mathfrak{h}$, also called η . Obviously $\eta \circ \bar{\bar{\xi}}|_V = \mathrm{Id}_V$, hence $\eta \circ \bar{\bar{\xi}} = \mathrm{Id}_{\mathfrak{h}}$, q.e.d □

5.7.3 COROLLARY *Let* $\mathfrak{n}$, Q *be such that any two elements of* R *are positively independent. Put* $\mathfrak{m} = \varinjlim \mathfrak{n}^C$. *Then* $\mathfrak{m} \to \mathfrak{n}$ *is a central extension, whose kernel is contained in* $\mathfrak{m}^{0}$, *and* $Q \ltimes \exp \mathfrak{m}$ *has a compact presentation.*

PROOF. Here the ground field is always K . The first two claims are proved in 4.7. For the last one we check the hypotheses of 5.6.1 . For the map $\pi : \mathfrak{m} \to \mathfrak{n}$ induced by the inclusions $\mathfrak{n}^C \to \mathfrak{n}$, the restrictions $\pi|\mathfrak{m}^{\alpha} : \mathfrak{m}^{\alpha} \to \mathfrak{n}^{\alpha}$ are isomorphisms for $0 \neq \alpha \in \mathrm{Hom}(Q,\mathbb{R})$, by 4.7. Therefore $\pi|\mathfrak{m}^C : \mathfrak{m}^C \simeq \mathfrak{n}^C$ for $C \in \mathcal{C}$, hence π induces an isomorphism $\varinjlim \mathfrak{m}^C \cong \varinjlim \mathfrak{n}^C = \mathfrak{m}$, i.e. $H_2(\mathfrak{m})^{0} = 0$ by 5.7.2. Also $\ker \pi \subset \mathfrak{m}'$ by 4.7, so $\mathfrak{m}^{ab} \cong \mathfrak{n}^{ab}$ and hence the sets R for $\mathfrak{m}$ and $\mathfrak{n}$ coincide, so $Q \ltimes \exp \mathfrak{m}$ has a compact presentation by 5.6.1 □

5.7.4 EXAMPLE We now give an example where $H_2(\mathfrak{n}|\mathbb{Q})^{\circ} \to H_2(\mathfrak{n})^{\circ}$ and $\varinjlim N^C \to N$ are not isomorphisms. This example shows by 4.7.3 that the map $H \to H_U$ of 4.5 is not an isomorphism in general and thus that it is necessary to pass from the abstract group H to the topological group H_U in order to obtain the desired isomorphism $H_U \cong Q \ltimes N$ if there is a compact presentation for the latter group.

The example will be the Lie algebra $\mathfrak{n}_2$ with Q-action defined below. Let $\mathfrak{n}_1$ be the free step 2 nilpotent Lie algebra over the p-adic field K on 3 generators X_1, X_2, X_3. Let Q be the free abelian group on two generators t_1, t_2. Let $\alpha_i : Q \to K^*$ be defined by $\alpha_i(t_j) = p^{\delta_{ij}}$, where δ_{ij} is the Kronecker delta and p is the characteristic of the residue field of K. There is a unique action of Q on $\mathfrak{n}_1$ by Lie algebra automorphisms over K such that

$$\mathfrak{n}_1^{\alpha_1} = K \cdot X_1$$

$$\mathfrak{n}_1^{\alpha_2} = K \cdot X_2$$

$$\mathfrak{n}_1^{-(\alpha_1+\alpha_2)} = K \cdot X_3 .$$

In the following diagram we see the plane $\mathrm{Hom}(Q,\mathbb{R})$ with basis $\beta_i = \nu_*(\alpha_i)$, $i = 1, 2$. The points $\gamma = a\beta_1 + b\beta_2$ which are in $\nu_* W(\mathfrak{n})$ are marked by dots. Next to $\gamma \in \nu_* W(\mathfrak{n})$ there is denoted a basis of $\mathfrak{n}_1^{\gamma}$.

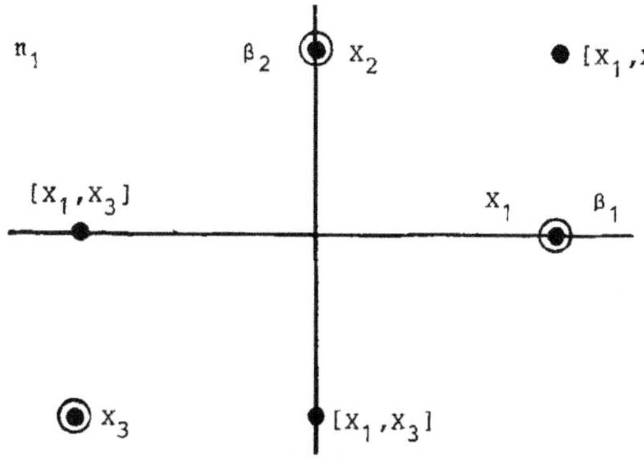

In the diagram, the points of $R = \nu_* W(\mathfrak{n}^{ab}) = \{\beta_1, \beta_2, -\beta_1 - \beta_2\}$ are encircled. So the first condition of our main theorem is satisfied: Any two elements of R are positively independent.

I claim that $\mathfrak{n}_2 := \varinjlim \mathfrak{n}_1^C$ is an example of the desired kind.

Claims For $\mathfrak{n}_2 = \varinjlim \mathfrak{n}_1^C$ the natural maps $H_2(\mathfrak{n}_2|\mathbb{Q})^O \to H_2(\mathfrak{n}_2)^O$ and $\varinjlim N_2^C \to N_2$ are not isomorphisms, but $Q \ltimes \exp \mathfrak{n}_2$ has a compact presentation.

The last claim follows from 5.7.3 and the diagram, since $\mathfrak{n}_1$ and $\mathfrak{n}_2$ have the same sets R, as $\ker(\mathfrak{n}_2 \to \mathfrak{n}_1) \subset \mathfrak{n}_2'$ by 4.7. The first two claims will follow from the following

Assertion $$\varinjlim \mathfrak{n}_1^C|\mathbb{Q} \to \varinjlim \mathfrak{n}_1^C$$
is not an isomorphism.

I first show that the assertion implies the first two claims. The map $\pi : \mathfrak{n}_2 = \varinjlim \mathfrak{n}_1^C \to \mathfrak{n}_1$ induced by the inclusions $\mathfrak{n}_1^C \to \mathfrak{n}_1$ restricts to isomorphisms of K-vector spaces $\mathfrak{n}_2^\alpha \to \mathfrak{n}_1^\alpha$ for $0 \neq \alpha \in \mathrm{Hom}(Q, \mathbb{R})$ by 4.7. Therefore $\varinjlim \mathfrak{n}_2^C|k \cong \varinjlim \mathfrak{n}_1^C|k$ for $\mathbb{Q} \subset k \subset K$. So the assertion implies $H_2(\mathfrak{n}_2|\mathbb{Q})^O \not\cong H_2(\mathfrak{n}_2)^O$ by 5.7.2, our first claim.

To prove the second claim, consider the Lie algebra $\varinjlim \mathfrak{n}_1^C|k$: It is nilpotent by 4.7, hence we have an abstract group $\exp(\varinjlim \mathfrak{n}_1^C|k)$ defined by Campbell-Hausdorff multiplication, see 2.5. The inclusions $N_2^C \to \exp(\varinjlim \mathfrak{n}_2^C|\mathbb{Q})$ induce the first arrow in

$$\varinjlim N_2^C \to \exp(\varinjlim \mathfrak{n}_2^C|\mathbb{Q}) \to \exp(\varinjlim \mathfrak{n}_2^C) .$$

Both homomorphisms are surjective, since their abelianizations are by 4.7, but the second one is not an isomorphism, by the assertion.

Proof of the assertion. The fact to be used here is that there is a non-zero $\mathbb{Q}$-linear map ρ from $K \otimes_{\mathbb{Q}} K$ to some $\mathbb{Q}$-vector space such that

$$\rho(\lambda \otimes \mu\nu) = \rho(\lambda\mu \otimes \nu) + \rho(\lambda\nu \otimes \mu)$$

for λ, μ, ν in K. Starting from such a map ρ we shall derive that $H_2(\mathfrak{n}_1|\mathbb{Q})^O \to H_2(\mathfrak{n}_1)^O$ is not an isomorphism - cf. the proof of 5.5.4 - , which proves the assertion by 5.7.2.

We first show the existence of such a map ρ. Let $x \mapsto x'$ be a non-zero derivation of K into itself [34]. Then

$$\rho : K \otimes_k K \to K$$

$$\rho(\lambda \otimes \mu) := \lambda \cdot \mu'$$

has the properties stated above. So it suffices to show that K has a non-zero derivation.

Let $k(T)$ be the field of functions in one indeterminate T over k. Given a derivation $k \to k$ of k and an element $a \in k(T)$, there is a unique derivation $k(T) \to k(T)$, $f \to f'$, extending the given derivation of k and such that $T' = a$. It follows by Zorn's lemma that every derivation of every field k extends to every purely transcendental extension field over k, and in more than one way, unless the transcendence degree is zero. On the other hand, given a separable algebraic extension field L of a field K, every derivation $K \to K$ of K extends to a unique derivation $L \to L$ of L. It follows that every field K of characteristic zero, which is not algebraic over $\mathbb{Q}$, admits a non-zero derivation. Our p-adic field K is not algebraic over $\mathbb{Q}$, since it is uncountable, whereas the field of all algebraic numbers is countable.

We now show that the natural map $H_2(\mathfrak{n}_1|\mathbb{Q})^{\circ} \to H_2(\mathfrak{n}_1)^{\circ}$ is not an isomorphism.

Let k be a field with $\mathbb{Q} \subset k \subset K$. The space $Z_2(k)^{\circ}$ (see 5.5.1) is spanned by the elements of the form

$$\lambda X_i \wedge \mu[X_{i+1}, X_{i+2}] , \quad \lambda, \mu \text{ in } K , \quad i = 1,2,3,$$

taking indices modulo 3. This follows from our picture of $\nu_* W(\mathfrak{n})$ two pages back. The space $B_2(k)^{\circ}$ is spanned by the set of elements

$$d_3(\lambda_1 X_1 \wedge \lambda_2 X_2 \wedge \lambda_3 X_3) =$$

$$- \lambda_1\lambda_2[X_1,X_2] \wedge \lambda_3 X_3 + \lambda_1\lambda_3[X_1,X_3] \wedge \lambda_2 X_2 - \lambda_2\lambda_3[X_2,X_3] \wedge \lambda_1 X_1 , \quad \lambda_i \in K .$$

For every i the k-linear map

$$m_i : K \otimes_k K \to \mathrm{span}\{\lambda X_i \wedge \mu[X_{i+1}, X_{i+2}]\}$$

$$\lambda \otimes \mu \longrightarrow \lambda X_i \wedge \mu[X_{i+1}, X_{i+2}]$$

is a k-linear isomorphism, as follows easily by decomposing $\mathfrak{n}_1 \wedge_k \mathfrak{n}$. Now let $\rho \neq 0$ be a k-linear map $K \otimes_k K \to K$ as above. There is a well defined k-linear map $\sigma : Z_2(\mathbb{Q})^{o} \to K$ such that

$$\sigma(\lambda X_i \wedge \mu[X_{i+1}, X_{i+2}]) = \rho(\lambda \otimes \mu) \quad \text{if} \quad i = 2$$

and $$\sigma(\lambda X_i \wedge \mu[X_{i+1}, X_{i+2}]) = \rho(\mu \otimes \lambda) \quad \text{if} \quad i = 1,3 .$$

The map σ vanishes on $B_2(\mathbb{Q})^{o}$ by the property of ρ .

This implies that the surjection $H_2(\mathfrak{n}_1|\mathbb{Q})^{o} \to H_2(\mathfrak{n}_1)^{o}$ is not an isomorphism. If it were an isomorphism, the natural map $\pi : Z_2(\mathbb{Q})^{o} \to H_2(\mathfrak{n}_1|\mathbb{Q})^{o}$ would have the bilinearity property $\pi(\lambda X \wedge \mu Y) = \pi(\lambda\mu X \wedge Y)$ for λ, μ in K . Therefore the composite map $\rho : K \otimes_{\mathbb{Q}} K \to K$ in the commutative diagram

$$\begin{array}{ccccc} K \otimes_{\mathbb{Q}} K & \xrightarrow{m_2} & Z_2(\mathbb{Q})^{o} & \xrightarrow{\pi} & H_2(\mathfrak{n}_1|\mathbb{Q})^{o} \\ & & & \searrow^{\sigma} & \downarrow \\ & & & & K \end{array}$$

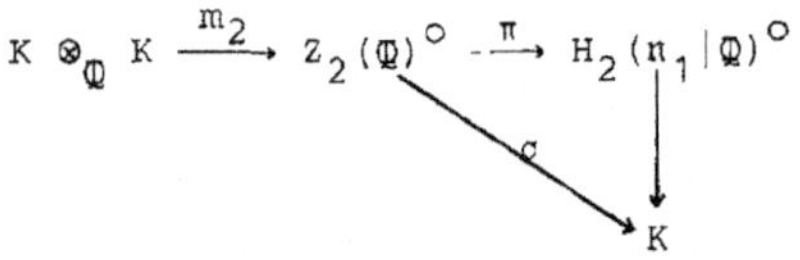

would be K-bilinear, i.e. $\rho(\lambda \otimes \mu) = \rho(\gamma\mu \otimes 1)$. But $\rho(\lambda \otimes 1 \cdot 1) = \rho(\lambda \otimes 1) + \rho(\lambda \otimes 1) = 0$, a contradiction.

5.7.5 Note that 5.7.2 and 5.7.3 again imply necessity of $H_2(\mathfrak{n})^{o} = 0$ for compact presentability, by 1.1.3 b), because the kernel of the map $Q \ltimes \exp(\varinjlim \mathfrak{n}^C) \to Q \ltimes \exp \mathfrak{n}$ is compactly generated as a normal subgroup only if it is zero.

5.7.6 Suppose any two elements of R are positively independent. I do not know whether $\varinjlim N^C \to \exp \varinjlim \mathfrak{n}^C|\mathbb{Q}$ is an isomorphism, not even whether the kernel of $\varinjlim N^C \to N$ is central. I also do not know whether

$$H_2(\mathfrak{n}) / \sum_{C \in \mathcal{C}} \mathrm{im}\Big(H_2(\mathfrak{n}^C) \to H_2(\mathfrak{n})\Big) \to H_2(\mathfrak{n})^{o}$$

is an isomorphism.

VI. S-ARITHMETIC GROUPS

In this chapter we apply our main result to S-arithmetic groups over number fields and algebraic groups over p-adic fields.

Let k be a field with $[k : \mathbb{Q}] < \infty$. A S-arithmetic group $G_{\mathcal{O}(G)}$ has a finite presentation iff G_{k_p} has a compact presentation, by a theorem of Kneser, see 6.1 for notations etc. Let K be a p-adic field. Let H be a maximal K-split solvable subgroup of G. Then the group G_K has a compact presentation iff H_K does, see 6.3.4. By our main result we can give necessary and sufficient conditions for H_K to have a compact presentation, if H is K-split solvable, see 6.2. In 6.3 a maximal K-split solvable subgroup H of a linear algebraic group G is described. In 6.4 necessary and sufficient conditions for compact presentability of G_K are given in case G is K-split, i.e. contains a maximal torus which is K-split. The proof is given in sections 6.5 and 6.6. The main result of 6.6 may be of independent interest. It gives a necessary and sufficient condition for the extendability of a representation of a Borel subgroup B of G to all of G, in case G is split reductive.

To sum up, in order to decide whether G_K has a compact presentation, one has to determine a maximal K-split solvable subgroup H of G, e.g. by the procedure in 6.3, and then to apply 6.2.3. In case G is K-split, one has the criterion 6.4.3, which should be easier to apply.

6.1 KNESER'S RESULT

In this section we recall Kneser's result which reduces the problem of finite presentability of S-arithmetic groups to the problem of compact presentability of certain locally compact groups.

6.1.1 In this section we shall use the following notations.

k - a finite extension field of $\mathbb{Q}$,

o - the ring of integers of k ,

p - a prime ideal of o ,

k_p - the completion of k with respect to the p-adic topology,

o_p - the ring of p-adic integers of k_p ,

S - a finite set of prime ideals of o ,

$o(S)$ - the ring of elements of k , which are p-integer for every $p \notin S$,

G - a linear algebraic group defined over k .

If $\rho_i : G \to Gl_{n_i}$, $i = 1,2$, are faithful representations of G defined over k , the inverse images of $Gl_{n_i,o(S)}$ with respect to $\rho_{i,k}$ are commensurable. Every group in the commensurability class of $(\rho_{i,k})^{-1} Gl_{n_i,o(S)}$ is called a *S-arithmetic group*. For $S = \emptyset$ a S-arithmetic group is called *arithmetic*.

We study the problem, which S-arithmetic groups are finitely presentable.

6.1.2 THEOREM (M. Kneser [36]) *A S-arithmetic subgroup of the linear algebraic group* G *is finitely generated (finitely presentable) iff* G_{k_p} *is compactly generated (compactly presentable) for every* $p \in S$.

The proof of the theorem makes use of the result of Borel and Harish-Chandra [19,15], that every arithmetic group has a finite presentation, which in turn is based on reduction theory.

Kneser's theorem has two advantageous features. Firstly, one has to deal with only one prime ideal $\mathfrak{p}$ at a time and secondly it relates problems about S-arithmetic groups to problems over local fields, where more structure is available. The problem, which groups $G_{k_\mathfrak{p}}$ are compactly generated, was solved by Borel and Tits ([22] Theorem 13.4, cf. 6.2.5). Making use of Kneser's result and results on reductive groups over local fields Behr proved the following theorem.

6.1.3 Theorem [9] *If* G *is reductive, every S-arithmetic subgroup of* G *has a finite presentation.*

A proof is given in 6.3.5.

6.2 The split solvable case

In this section we deal with the case that our group is split solvable over the local field K . Theorem 6.2.3 is a corollary of our main theorem 5.6.1. We obtain as a corollary (6.2.4) the criterion for finite presentability.

Facts about algebraic groups are taken from Borel, Tits [22].

6.2.1 Let K be a field of characteristic zero. An algebraic subgroup of Gl_n defined over K is called *trigonalisable over* K if it is conjugate over K to a subgroup of the group B_n of all upper triangular matrices. A connected solvable linear algebraic group H defined over K is called K-*split solvable* if one of the following equivalent conditions holds.

1) H *is* K-*isomorphic to a matrix group which is trigonalisable over* K.
2) H *contains a maximal torus which is* K-*split.*
3) *Every image of* H *under a* K-*morphism* $G \to Gl_m$ *is trigonalisable over* K .

4) *H is K-solvable, i.e. there is a sequence* $H = H_o \supset H_1 \supset \ldots \supset H_t = \{e\}$ *of connected K-subgroups with* H_{i+1} *normal in* H_i *and all quotients* H_i/H_{i+1} *K-isomorphic to* $\mathbb{G}_a$ *or* $\mathbb{G}_m = GL_1$.

6.2.2 Let H be a K-trigonalisable group. Let U be its unipotent radical. It is defined over K . Let $\mathfrak{u}$ be the Lie algebra of U . The group G/U = T is a torus defined and split over K . The adjoint representation of H on $\mathfrak{u}$ induces a K-representation of H on the homology of $\mathfrak{u}$, whose kernel contains U , so induces a K-representation of T on $H_i(\mathfrak{u})$. Let $X^*(T)$ be the *character group* of T , i.e. the abelian group of morphism from T to $\mathbb{G}_m$. Every morphism $T \to \mathbb{G}_m$ is defined over K , since T is K-split. For every K-representation $\rho : T \to Gl(V)$ let $W(\rho) \subset X^*(T)$ be the set of its *weights*. So $\chi \in X^*(T)$ belongs to $W(\rho)$ iff there is a non-zero vector $v \in V$ such that $\rho(t)v = \chi(t)v$ for every $t \in T$.

6.2.3 THEOREM *Let K be a p-adic field. Let H be a K-split solvable linear algebraic group. The group* H_K *of its K-points is a locally compact topological group with respect to the topology given by K . The group* H_K *has a compact presentation iff the following two conditions hold.*

1) *Any two weights of the representation of T on* $H_1(\mathfrak{u}) = \mathfrak{u}^{ab}$ *are positively independent in* $X^*(T) \otimes_{\mathbb{Z}} \mathbb{R}$.

2) *Zero is not a weight of the representation of T on* $H_2(\mathfrak{u})$.

Since every linear algebraic group G over a local field K of characteristic zero contains a maximal K-split subgroup H and G_K/H_K is compact ([22] Theorem 9.3), 6.2.3 and 6.1.2 imply

6.2.4 COROLLARY *The S-arithmetic subgroup of the linear algebraic group G over k has a finite presentation iff for every* $p \in S$ *a maximal* k_p*-split subgroup H of G fulfills conditions 1) and 2) of 6.2.3* □

Compare this with the following special case of a theorem of Borel and Tits [22] Theorem 13.4.

6.2.5 THEOREM *Let* K *be a p-adic field. Let* G *be a* K*-trigonalisable linear algebraic group. The group* G_K *of its* K*-points has a compact set of generators iff zero is not a weight of the representation of* $T = G/U$ *on* $H_1(\mathfrak{u})$ *induced by the adjoint representation.*

6.2.6 REMARK Note that the Lie algebra $\mathfrak{u}$ is defined over K . So in the theorems we may replace $H_i(\mathfrak{u})$ by $H_i(\mathfrak{u})_K = H_i(\mathfrak{u}_K)$ and T by T_K , since T is K-split. Here we consider $\mathfrak{u}$ and T as algebraic sets with coefficients in some "big" field.

PROOF OF 6.2.3 T is a k-split torus. So there is a discrete free abelian subgroup Q of T_K with rank Q = dim T , such that T_K/Q is compact and hence the map

$$X^*(T) \otimes_{\mathbb{Z}} \mathbb{R} \to \operatorname{Hom}(Q,\mathbb{R})$$

$$\lambda \otimes 1 \to \nu_*(\lambda | Q)$$

is an isomorphism of real vector spaces. For the definition of ν_* see 5.6. Let now S be a maximal K-split subtorus of G . The natural map $G \to G/U$ induces an isomorphism $\pi : S \to T$. The exponential map of matrices $\operatorname{Exp}(z) = 1 + z + \frac{z^2}{2!} + \ldots$ induces an isomorphism of the Lie group corresponding to $\mathfrak{u}_K$ with U_K . Now apply 5.6 to $\pi^{-1}(Q) \ltimes U_K$ using remark 6.2.5 and the fact that

$$X^*(T) \to X^*(T) \otimes_{\mathbb{Z}} \mathbb{R}$$

is injective □

Similarly 6.2.5 follows from 3.2.2 □

6.3 A MAXIMAL SPLIT SOLVABLE SUBGROUP

In this section we describe a maximal split solvable subgroup of a linear algebraic group G (6.3.3).

6.3.1 Let G be a connected reductive linear algebraic group defined over a field k of characteristic zero. Let T_1 be a maximal k-split subtorus of G and let T_2 be a maximal subtorus of G containing T_1. Let $\Phi(T_i,G)$, $i = 1,2$, be the set of *roots* of T_i on G, i.e. the set of weights of the representation of T_i on the Lie algebra $\mathfrak{g}$ of G given by the adjoint representation. For every $b \in \Phi(T_2,G)$ there is a corresponding radicial subgroup U_b of G defined by the existence of an isomorphic $\vartheta_b : \mathbb{G}_a \to U_b$ such that

$$t\vartheta_b(x)t^{-1} = \vartheta_b(b(t)\cdot x),\ t \in T_2,\ x \in \mathbb{G}_a .$$

For a given partial order on the group $X^*(T_1)$ of (k-)morphisms $T_1 \to \mathbb{G}_m$ let $\Psi = \{b \in \Phi(T_1,G)\ ;\ b > 0\}$. Suppose the order is chosen in such a way that $\Phi(T_1,G)\setminus\{0\} = \Psi \dot{\cup} -\Psi$, which is obviously possible. Let U_Ψ be the group generated by the set of U_b, where $b \in \Phi(T_2,G)$ and the restriction of b to T_1 belongs to Ψ. The group U_Ψ does not change if one replaces T_2 by another maximal torus containing T_1 ([22] 3.8(ii)).

6.3.2 LEMMA *$T_1 \cdot U_\Psi$ is a maximal k-trigonalizable subgroup of G. The group U_Ψ is a normalized by the centralizer $Z(T_1)$ of T_1, in particular by T_2.*

PROOF. U_Ψ is unipotent [22] 3.8 (iv), and normalized by $Z(T_1)$, [22] 3.8 (ii).

Let A be a maximal k-trigonalizable subgroup of G containing $T_1 \cdot U_\Psi$. We have $A = T_1 \cdot U$, where U is the unipotent radical of A and U is defined over k. Then there is an order on $X^*(T_1)$ such that the weights of T_1 on U are > 0 ([22] 3.6). It follows that

$\Phi(T_1,U) = \Psi$, since $\Phi(T_1,G)\setminus\{0\} = \Psi \dot{\cup} -\Psi$, $\Psi \subset \Phi(T_1,U)$ and $\Phi(T_1,U) \cap -\Phi(T_1,U) = \emptyset$. This proves the lemma □

Having determined a maximal k-trigonalizable subgroup of G for reductive G , it is easy to do so for general G . Let G be a linear algebraic group defined over k and let L be a maximal connected reductive subgroup of G defined over k . Let T_1 be a maximal k-split subtorus of L . For an order on $X^*(T_1)$ define $\Psi = \{b \in \Phi(T_1,L) ; b>0\}$.

6.3.3 COROLLARY $H = T_1 \cdot U_\Psi \cdot \mathrm{Rad}_u(G)$ *is a maximal k-trigonalizable subgroup of* G □

6.3.4 If K is a local field of characteristic zero, and H is a maximal K-trigonalizable subgroup of G , then G_K/H_K is compact, [22] Prop. 9.3, hence G_K has a compact presentation iff H_K does, by 1.1.2. So we can apply the theorem of 6.2 to H .

6.3.5 We obtain Behr's theorem [9], see 6.1.3, as a corollary: If G is reductive, $\mathrm{Rad}_u G = \{e\}$ and $\mathrm{Rad}_u(H) = U_\Psi$, so the weights of T_1 on $H_1(u)$ and on $H_2(u)$ form a subset of $\{b \in X^*(T_1) ; a > 0\}$ and the conditions 1) and 2) of theorem 6.2.3 hold. Actually, it is easy to see that there is an element in $T_{1,K}$ that acts contracting on U_Ψ , so the much easier theorem 1.3.1 (or 4.1.3) can be applied □

6.4 G SPLIT

In this section we give necessary and sufficient conditions for G_K to have a compact presentation, supposed G is K-*split* (not necessarily solvable) i.e. that G contains a maximal torus which is K-split (theorem 6.4.3). These conditions are formulated in terms of the structure of G itself, in contrast with 6.2 where the conditions were given in terms of a maximal K-split solvable subgroup. It would be interesting to have a generalization of theorem 6.4.3 to non-split groups.

6.4.1 In order to state the result we have to recall some more facts and notations (see [22]). Let L be a connected reductive linear algebraic group defined over a field k of characteristic zero. Suppose there is a maximal subtorus T of L defined and split over k. Let Φ be the set of *roots* of T on L. The set Φ forms a *root system* in $X^*(T) \otimes_{\mathbb{Z}} \mathbb{R}$ in the sense of [22], so Φ does not necessarily span $X^*(T) \otimes \mathbb{R}$. Let $(\, , \,)$ be a positive definite scalar product on $X^*(T) \otimes \mathbb{R}$, which is admissible, i.e. invariant under the Weyl group of Φ. Choose a *Weyl chamber* C, i.e. connected component of the complement in $X^*(T) \otimes \mathbb{R}$ of the union of the hyperplanes orthogonal to the roots. A root is called *positive*, if $(\alpha,x) > 0$ for every $x \in C$. The set of positive roots is denoted Φ_+. A root is called *simple*, if it is positive and not the sum of two positive roots. The set of simple roots (with respect to a Weyl chamber C) is denoted Δ.

Let N_+ and N_- be the subgroup of L generated by the radicial subgroups U_a, $a \in \Phi_+$ ($-a \in \Phi_+$ resp.). The groups N_+ and N_- are unipotent algebraic groups defined over k. The *Borel subgroup* B_+ and B_- resp. corresponding to C ($-C$ resp.) is $T \cdot N_+$ ($T \cdot N_-$ resp.).

Every k-representation of a k-split reductive group L is the direct sum of absolutely irreducible k-representations. Let ρ be an irreducible k-representation of L on V. There is a unique line (= one-dimensional vector subspace) l_ρ of V kept invariant by B_-. It is a weight space of T, the corresponding weight in $X^*(T)$ is called the *dominant weight* of ρ.

For a - not necessarily irreducible - k-representation of L on a vector space V, the set of dominant weights of the irreducible k-subrepresentations of ρ will be called the *set of dominant weights* of ρ or the set of dominant weights of the L-module V and will be denoted $W_{dom}(\rho)$ or $W_{dom}(V)$.

6.4.2 We are now ready to state the main theorem of this section. Let K be a p-adic field. Let G be a connected linear algebraic group defined over K . Let U be its unipotent radical. Let $\mathfrak{u}$ be the Lie algebra of U . The adjoint action of G on $\mathfrak{u}$ induces a K-representation of $L := G/H$ on $H_i(\mathfrak{u})$. We look at the topology of G_K given by the topology of the locally compact field K .

6.4.3 THEOREM *Suppose* L *is K-split reductive. Then* G_K *has a compact presentation iff the following two conditions hold.*

1) Any two elements of $\Delta \cup - W_{dom}(H_1(\mathfrak{u}))$ *are positively independent in* $X^*(T) \otimes_{\mathbb{Z}} \mathbb{R}$.

2) $0 \notin W_{dom}(H_2(\mathfrak{u}))$.

A proof of this theorem will be given in sections 6.5 and 6.6.

Compare 6.4.3 with the following special case of a theorem of Borel and Tits ([22] 13.4).

6.4.4 THEOREM *Suppose* L *is k-split reductive. Then* G_K *has a compact set of generators iff*

$$0 \notin W_{dom}(H_1(\mathfrak{u})) .$$

6.4.5 REMARK Let L be a reductive group defined over k with a maximal torus T which is k-split. A representation ρ of L on V defined over k has the property

1) Any two elements of $\Delta \cup - W_{dom}(V)$ *are positively independent*

iff the following two conditions hold.

1a) Any two (not necessarily different) elements of $W_{dom}(V) \cap \Phi^{\perp}$ *are positively independent.*

1b) There is no element in $W_{dom}(V)$ *, which is a positive multiple of some simple root.*

PROOF. Obviously, 1) ⇒ 1a) and 1b). To see the converse, note first that $0 \notin W_{dom}(V)$ by 1a). Any two elements of Δ , actually of Φ^+ , are positively independent. So there are two types of pairs of elements of $\Delta \cup -W_{dom}(V)$ left to check, namely those containing exactly one simple root, which are taken care of by 1b), and pairs of elements (λ_1, λ_2) in $W_{dom}(V)$. A convex combination of non-zero dominant weights λ_1, λ_2 can be zero only if λ_1 and λ_2 are both in $\Delta^\perp = \Phi^\perp$, since $(\lambda_i, \alpha) \geq 0$ for every $\alpha \in \Delta$ □

The irreducible representations corresponding to weights of the type considered in 1a) are exactly those whose kernel contains the semisimple part L' of L .

If a dominant weight $\lambda \neq 0$ is a positive multiple of a simple root α say, then α is an isolated point in the Witt-Dynkin-diagram of Δ , because $(\alpha, \beta) \leq 0$ for any two different simple roots and $(\beta, \lambda) \geq 0$ for every dominant weight λ and $\beta \in \Delta$. So 1b) is equivalent to

1b') If W *is an irreducible submodule of* V *the image of the corresponding representation* $\rho W : L \to Gl(W)$ *is not a simple group of rank one.*

Hence if L has no simple factors of rank one, 1) is equivalent to 1a). On the other hand, if G is a linear algebraic group defined over the *p*-adic field K such that G modulo its unipotent radical U is split simple of rank one, then G_K has no compact presentation, unless $U = \{e\}$.

6.4.6 REMARK Under the hypotheses of theorem 6.4.3, condition 2) is equivalent to

2') Let $f : H \to G$ *be a surjective morphism of linear algebraic groups over* K *whose kernel is central and contained in the commutator subgroup of the unipotent radical of* H *. Then* f *is an isomorphism.*

PROOF. Associating with a unipotent algebraic group its Lie algebra defines an quivalence of the category of unipotent algebraic groups over K with the category of nilpotent Lie algebras over K , if char $K = 0$ (see [26]). So our claims follows from the Hopf description of $H_2(\mathfrak{u})$, see 5.2 □

6.5 PROOF OF THEOREM 6.4.3

6.5.1 Let G be a linear algebraic group defined over K , let U be its unipotent radical. Suppose $L := G/U$ contains a maximal torus T which is K-split. Let $\pi : G \to L$ be the natural map. There is a maximal reductive subgroup R of G defined over K containing a maximal torus T which is K-split. The morphism $R \to L$ induced by π is an isomorphism. Let G^o be the connected component of the identity in G with respect to the Zariski topology. Since G^o_K is open and of finite index in G_K , we may assume that G is connected (with respect to the Zariski topology). So G is the semidirect product of R with U . A maximal K-trigonalizable subgroup of G is $H = B_+ \cdot U = T \cdot N_+ \cdot U$, where N_+ is the subgroup of R generated by the radicial subgroups U_a , $a \in \Phi_+$. This follows from 6.3.3 by looking at a partial order on $X^*(T) \otimes \mathbb{R}$ such that $\Phi_+ = \{a \in \Phi \; ; \; a > 0\}$. The unipotent radical $\mathrm{Rad}_u(H)$ of H is

$$\mathrm{Rad}_u(H) = N_+ \cdot U \quad .$$

For the corresponding Lie algebras we have

$$\mathfrak{r}_u = \mathfrak{n}_+ \ltimes \mathfrak{u} \quad .$$

For a vector space V defined over K with a K-representation of T let W(V) be the set of weights of T on V .

6.5.2 Lemma $W(H_1(r_u)) = \Delta \dot{\cup} - W_{dom}(H_1(u))$.

Here $\dot{\cup}$ denotes disjoint union.

The lemma implies that condition 1) of 6.2.3 for r_u is equivalent to condition 1) of 6.4.3.

Proof of 6.5.2 For a representation ρ of a Lie algebra $\mathfrak{g}$ on a vector space V one defines $H_o(\mathfrak{g};V)$ as the quotient of V by the subspace generated by the set of elements $\rho(X)v$, $X \in \mathfrak{g}$, $v \in V$.

The commutator algebra r_u' of r_u is

$$r_u' = n_+' \ltimes (u' + [n_+,u] \ .$$

So we have an exact sequence

$$0 \to H_o(n_+,H_1(u)) \to H_1(r_u) \to H_1(n_+) \to 0 \quad ,$$

hence $W(H_1(r_u)) = W(H_1(n_+)) \cup W(H_o(n_+,H_1(u)))$.

For a,b in Φ_+ not proportional, (U_1,U_b) is the subgroup $U_{(a,b)}$ generated by the groups U_c , where c runs through the set (a,b) of all roots of the form $ia + jb$, $i,j > 0$ ([22] 2.5 Remarque). This implies $\Delta = W(H_1(n_+))$, since Φ is a reduced root system: ϕ is identical with the root system over the algebraic closure, hence reduced, cf. [18, 14.8].

Let V be a vector space defined over K with a K-representation of L . The lemma follows if we have shown that

$$W(H_o(n_+;V)) = -W_{dom}(V) \quad .$$

We may assume that the representation of L on V is irreducible. Let ℓ_ρ be the unique line of V kept invariant by B_+ , so ℓ_ρ is the dominant weight space with respect to the Weyl chamber $-C$. Its weight in $X^*(T)$ is the negative of the dominant weight with respect to the Weyl chamber C , as is seen by applying the element of the Weyl group mapping C to $-C$. It is a standard fact about representations of semi-simple Lie algebras that V is generated by ℓ_ρ as a module over the Lie algebra $\mathfrak{n}_+$, so the natural map $\ell_\rho \to H_o(\mathfrak{n}_+;V)$ is surjective. On the other hand, $H_o(\mathfrak{n}_+;V)$ is non-zero, since $\mathfrak{n}_+$ acts nilpotently on V □

6.4.3 follows from the results in 6.2 and 6.3 once we have shown the following lemma.

6.5.3 LEMMA *Suppose* L *is K-split. The map* $H_2(\mathfrak{u}) \to H_2(\mathfrak{r}_u)$ *given by the inclusion* $\mathfrak{u} \to \mathfrak{r}_u$ *induces an isomorphism of the subspace of* L-*fixed elements of* $H_2(\mathfrak{u})$ *to the subspace of* T-*fixed points of* $H_2(\mathfrak{r}_u)$.

Note that an irreducible K-representation of L is trivial iff its dominant weight is zero.

Let us call a Lie algebra $\mathfrak{g}$ defined over K with a K-representation ρ of L on $\mathfrak{g}$ such that $\rho(x)$ is a Lie algebra automorphism defined over K for very $x \in L$ a *Lie algebra with* L-*action* for short. A morphism of Lie algebras with L-action is a map that is compatible with the two defining structures, i.e. it is a Lie algebra homomorphism defined over K and it is L-equivariant.

We now start the proof of 6.5.3. In this section we reduce it to the main result of the next section, theorem 6.6.2. We actually only need its corollary 6.6.9.

If a group G acts on a set X we denote by X^G the set of G-fixed points.

We start out with the easy part $\dim H_2(\mathfrak{r}_u)^T \geq \dim H_2(\mathfrak{u})^L$. There is an exact sequence of Lie algebras with L-action

$$(1) \qquad H_2(\mathfrak{u})^L \rightarrowtail \mathfrak{a} \twoheadrightarrow \mathfrak{u}$$

such that $H_2(\mathfrak{u})^L$ is central in $\mathfrak{a}$, contained in the commutator subalgebra of $\mathfrak{a}$ and contained in $\mathfrak{a}^L$. This is seen by a Hopf type construction starting from a free Lie algebra $\mathfrak{h}$ over a L-module $V \subset \mathfrak{u}$, such that $V \to \mathfrak{u}^{ab}$ is an isomorphism of L-modules (cf. 5.2.2, proof of 5.2.5). Similarly there is an exact sequence of Lie algebras with T-action

(2) $$H_2(r_u)^T \rightarrowtail h \xrightarrow{\Psi} r_u$$

such that $H_2(r_u)^T \subset h' \cap h^T \cap$ center of h. Form the semidirect product of a with n_+ to obtain from (1) a central extension of Lie algebras with T-action

$$H_2(u)^L \rightarrowtail a \ltimes n_+ \twoheadrightarrow u \ltimes n_+$$

and apply the universal property of the Hopf extension to obtain a surjection $h \to a \ltimes n_+$ over the identity of r_u, hence a natural surjection $H_2(r_u)^T \to H_2(u)^L$.

We now show that the natural map $H_2(u)^L \to H_2(r_u)^T$ is surjective thus finishing the proof. Look at the central extension (2) of r_u. Since every K-representation of the K-split torus T is completely reducible, we have $h = \oplus h^\sigma$, $r_u = \oplus r_u^\sigma$, $\sigma \in X^*(T)$. Since Ψ is a morphism of T-modules and $\ker \Psi \subset h^0 = h^T$, the maps $h^\sigma \to r_u^\sigma$ are isomorphisms for $\sigma \neq 0$. So there is a unique section $\alpha : \bigoplus_{\sigma \neq 0} r_u^\sigma \to \bigoplus_{\sigma \neq 0} a^\sigma$ for Ψ which is T-equivariant. So for $X \in r_u^\sigma$, $\sigma \neq 0$, the element $\alpha(X)$ is the unique element X' of h^σ such that $\Psi(X') = X$. Since Ψ is a Lie algebra homomorphism, the same holds for α, as far as that makes sense, i.e.

$$[\alpha(X), \alpha(Y)] = \alpha[X,Y]$$

if X,Y and $[X,Y]$ are in $\mathrm{Def}(\alpha) = \bigoplus_{\sigma \neq 0} r_u^\sigma$. In particular, $\alpha | n_+$ is a Lie algebra homomorphism. Let $v = \Psi^{-1}(u)$. Then

(3) $$H_2(r_u)^T \rightarrowtail v \twoheadrightarrow u$$

is a central extension of u. We want (3) to be an exact sequence of Lie algebras with R-action and $H_2(r_u)T \subset v'$ so that we get a map from (1) to (3) and thereby the desired surjection $H_2(u)^L \to H_2(r_u)^T$.

Both these properties of (3) follow from the main result of the next section 6.6, actually from its corollary 6.6.9. We have a representation φ of $\mathfrak{n}_+$ on $\mathfrak{v}$ given by $\alpha|\mathfrak{n}_+$ followed by the representation of $\alpha(\mathfrak{n}_+)$ on $\mathfrak{v}$ given by the adjoint representation ad of $\mathfrak{h}$. We also have a representation of T on $\mathfrak{v}$. These two are compatible, i.e. $\varphi(\mathfrak{n}^\alpha)\mathfrak{v}^\beta \subset \mathfrak{v}^{\alpha+\beta}$ for $\alpha \in \Phi_+$, $\beta \in X^*(T)$. So there is a unique K-representation, also called φ, of the Borel subgroup $B_L := L \cap \pi^{-1}(B)$ of L on $\mathfrak{v}$ compatible with these two representations. There is a representation of R on $\mathfrak{u}$, given by the adjoint representation of G. The map $\Psi|\mathfrak{v} : \mathfrak{v} \to \mathfrak{u}$ in (3) is a homomorphism of B-modules whose kernel consists of T-fixed vectors. So by 6.6.9 there is a unique K-representation of L on $\mathfrak{v}$ extending the representation φ of B on $\mathfrak{v}$, the map Ψ is a homomorphism of R-modules, $\ker \Psi = H_2(\mathfrak{r}_u)^T \subset \mathfrak{v}^R$ and there is a complementary R-submodule $\mathfrak{w}$ of $\ker \Psi$ in $\mathfrak{v}$. So (3) is an exact sequence of Lie algebras with R-action.

Furthermore, $H_2(\mathfrak{r}_u)^T \subset \mathfrak{r}_u' = (\ker \Psi \oplus \mathfrak{w} \oplus \alpha(\mathfrak{n}_+))' = \mathfrak{w}' + [\alpha(\mathfrak{n}_+), \mathfrak{w}] + \alpha(\mathfrak{n}_+)'$. The second summand is contained in $\mathfrak{w}$, the third one in $\alpha(\mathfrak{u}_+)$, hence $\ker \Psi = H_2(\mathfrak{r}_u)^T \subset \mathfrak{w}' \subset \mathfrak{v}'$. So (3) induces an isomorphism $\mathfrak{v}^{ab} \xrightarrow{\sim} \mathfrak{u}^{ab}$. Hence there is a surjective map of exact sequences with R-action - note that $\pi|R : R \xrightarrow{\sim} L$ - from (1) to (3), in particular $H_2(\mathfrak{u})^L \to H_2(\mathfrak{r}_u)^T$ is surjective, q.e.d. Actually it is easy to see that the two maps $H_2(\mathfrak{u})^L \rightleftarrows H_2(\mathfrak{r}_u)^T$ defined in the two parts of the proof are inverse of each other □

6.6 Extending a representation of a Borel subgroup

Let R be a k-split reductive group, k a field of characteristic zero. We use the notations of 6.4.1. The proposition to be proved in this section gives a necessary and sufficient condition for a finite dimensional representation of a Borel subgroup B of R to extend to all of R.

6.6.1 Let V be a vector space defined over k and let φ be a k-representation of B on V. Then V decomposes into weight spaces

(1) $$V = \oplus V^{\beta} \ , \ \beta \in X^*(T) \ .$$

Set $W(V) = \{\beta \in X^*(T) \ ; \ V^{\beta} \neq 0\}$. The representation of N on V induces a representation $d\varphi$ of its Lie algebra $\mathfrak{n}$ on V. For every root $\alpha \in \Phi$ let $\alpha^{\vee}$ be the *inverse root*, i.e. the linear form $X^*(T) \otimes \mathbb{R} \to \mathbb{R}$ vanishing on the orthogonal complement of α and such that $\langle\alpha,\alpha^{\vee}\rangle = 2$. So the *reflection* σ_{α} of Φ corresponding to α is

(2) $$\sigma_{\alpha}(x) = x - \langle\alpha,\alpha^{\vee}\rangle\, \alpha$$

for $x \in X^*(T) \otimes \mathbb{R}$. According to our notations, $\mathfrak{r}^{\alpha}$ denotes the weight space in the Lie algebra $\mathfrak{r}$ of R corresponding to the root $\alpha \in \Phi$.

6.6.2 THEOREM *A k-representation φ of B on V extends to a k-representation of R on V iff the following two conditions hold.*

1) For every $\beta \in W(V)$ and $\alpha \in \Phi$ we have

$$\langle\beta,\alpha^{\vee}\rangle \in \mathbb{Z} \ .$$

2) For every $\beta \in W(V)$, $\alpha \in \Phi_+$ such that $\langle\beta,\alpha^{\vee}\rangle \geq 0$ and every non zero $X_{\alpha} \in \mathfrak{r}^{\alpha}$ the map

$$X_{\alpha}^{\langle\beta,\alpha^{\vee}\rangle} : V^{\sigma_{\alpha}(\beta)} \to V^{\beta}$$

is a linear isomorphism.

There is only one representation of R on V extending the given representation φ of B on V.

PROOF. We first study the case $R = Sl_2$.

6.6.3 Recall the following basic facts about finite dimensional representations of Sl_2 (cf. [24] VIII § 1). The Lie algebra sl_2 of Sl_2 has the basis

$$x = \begin{pmatrix} 0 & 1 \\ 0 & 0 \end{pmatrix}, \ y = \begin{pmatrix} 0 & 0 \\ 1 & 0 \end{pmatrix}, \ h = \begin{pmatrix} 1 & 0 \\ 0 & -1 \end{pmatrix} \ .$$

The following relations hold

$$[h,x] = 2x \ , \ [h,y] = -2y \ , \ [x,y] = h \ . \qquad (3)$$

Every finite dimensional representation of $\mathfrak{sl}_2$ is completely reducible and for every $n \in \mathbb{N}_0$ there is exactly one (absolutely) irreducible representation V_n of dimension $n+1$, which can be described as follows. V_n is the direct sum of the one-dimensional weight spaces $V_n^{-n}, V_n^{-n+2}, \ldots, V_n^{+n}$, where

$$hv = iv \quad \text{for } v \in V_n^i \text{ and } i = -n, -n+2, \ldots, +n$$

$$x : V_n^i \overset{\sim}{\to} V_n^{i+2} \text{ for } i \neq n \ , \ -n-2$$

$$y : V_n^i \overset{\sim}{\to} V_n^{i-2} \text{ for } i \neq -n, \ n+2 \quad .$$

For every finite dimensional $\mathfrak{sl}_2$-module V and $i \in \mathbb{Z}$ define

$$V^i = \{v \in V \, ; \, hv = iv\} \ .$$

Then

$$V = \bigoplus_{i \in \mathbb{Z}} V^i \qquad (4)$$

and

$$x^i : V^{-i} \to V^{+i} \ , \ y^i : V^i \to V^{-i} \qquad (5)$$

are linear isomorphisms.

In Sl_2 the subgroup of diagonal elements is a maximal $\mathbb{Q}$-split torus. The element x spans a root space corresponding to the root α say. Then y spans the root space corresponding to $-\alpha$. We have $\sigma_\alpha(v) = -v$. Let the ray spanned by α be the Weyl chamber C. The exponential of the one-dimensional subspace spanned by x resp. y is N resp. N_-. So B_+ resp. B_- consists of all upper (lower) triangular matrices of determinant one.

For every representation ρ of $\mathfrak{sl}_2$ on a finite dimensional vector space V there is a unique representation π of Sl_2 on V, such that π and ρ are compatible, i.e. such that $\pi \exp X = \exp \rho(X)$ for every nilpotent $X \in \mathfrak{sl}_2$. The subspace V^i is the weight space of the weight $\frac{i}{2}\alpha$. The representation ρ is defined over k iff π is defined over k.

6.6.4 Having recalled these facts we can now prove the proposition for $R = Sl^2$. Necessity of the two conditions follows from (4) and (5). We prove sufficiency by induction on $\dim V$.

The group $X^*(T)$ is an infinite cyclic group with generator $\frac{\alpha}{2}$. We set $V^i := V^{i\cdot\frac{\alpha}{2}}$ for $i \in \mathbb{Z}$. We have $V = \bigoplus_{i\in\mathbb{Z}} V^i$. This follows from condition 1), but also from equation (1). Let i_o be maximal such that $V^{i_o} \neq 0$. Condition 2) implies that $\dim V^i = \dim V^{-i}$ for $i \in \mathbb{Z}$. Let W be the B-submodule of V generated by V^{-i_o}, so

$$W = \bigoplus_{j=0}^{i_o} x^j V^{-i_o}, \tag{6}$$

since $x^{j+1} V^{-i_o} \subset V^{i_o+2} = 0$. So conditions 1) and 2) are satisfied for W as well. If there is a representation of Sl_2 on W extending the given representation of B on W, we have for the corresponding representation of sl_2 on W the following formula

$$y x^j v = j(i_o - j+1)x^{j-1} v \tag{7}$$

for $v \in V^{-i_o} = W^{-i_o}$ and $j \in \mathbb{N}_o$, as is easily seen by induction on j. Conversely, since $x^j : W^{i_o} \to W^{-i_o+2j}$ is an isomorphism for $j = 0,1,\dots,i_o$, we can associate to y a linear map $W \to W$ by formula (7) which is easily seen to define a representation of sl_2 on W. This implies that there is a k-representation π of $R = Sl_2$ on W extending the given representation of B on W.

There is a complementary submodule U of W in V defined over k, namely $U = \bigoplus_{i\in\mathbb{Z}} (V^{-i_o+2i} \cap \ker x^{i_o-i}) \oplus \bigoplus_{i \not\equiv i_o \bmod 2} V^i$ satisfying conditions 1) and 2). So the existence part of proposition 6.6.2 follows for $R = Sl_2$ by induction on $\dim W$.

6.6.5 The uniqueness part is shown by the usual trick. We first prove that a point of a representation space V of R is fixed by $R = Sl_2$ iff it is fixed by B. It suffices to show this for irreducible representations of R, since every representation of R is completely reducible. If V is an irreducible Sl_2-module, hence an irreducible $\mathfrak{sl}_2$-module containing a non-zero B-fixed element $v \in V^o \cap \ker x$, then in the notation of (6) above we have $i_o = 0$ and hence $yv = 0$ by (7), so v is $\mathfrak{sl}_2$-fixed, hence Sl_2-fixed and spans V, since V is irreducible.

If now V_1 and V_2 are two R-modules, the vector space $\operatorname{Hom}(V_1,V_2)$ is a R-module in a natural way. Its R-(B-)fixed elements are the R-(B-)module homomorphisms. So the fact just shown implies, that every B-module homomorphism between R-modules is actually an R-module homomorphism. If we apply this to the identity map of V endowed with two representations of R extending the given representation of B we get the uniqueness statement. Similarly for modules over the respective Lie algebras.

6.6.6 R GENERAL

Let $\mathfrak{t}$ be the Lie algebra of T. For every $\alpha \in \Phi$ there is a linear form $\overline{\alpha} \in \mathfrak{t}^*$ on $\mathfrak{t}$ such that

$$[h,x] = \overline{\alpha}(h)x \quad \text{for every} \quad h \in \mathfrak{t}, x \in \mathfrak{r}^\alpha . \tag{8}$$

Let $\alpha \in \Phi_+$. Define S_α as the subgroup of R generated by the radicial subgroups U_α and $U_{-\alpha}$. The Lie algebra of S_α is $\mathfrak{s}_\alpha = \mathfrak{r}^\alpha \oplus \mathfrak{r}^{-\alpha} \oplus [\mathfrak{r}^\alpha,\mathfrak{r}^{-\alpha}]$. Set $\mathfrak{h}_\alpha = [\mathfrak{r}^\alpha,\mathfrak{r}^{-\alpha}]$. There is a k-Lie algebra isomorphism $\vartheta_\alpha : \mathfrak{sl}_2 \to \mathfrak{s}_\alpha$ such that $\vartheta_\alpha(x) \in \mathfrak{r}^\alpha$, $\vartheta_\alpha(y) \in \mathfrak{r}^{-\alpha}$ and hence $\vartheta_\alpha(h) \in \mathfrak{h}_\alpha$. Note that, while ϑ_α is not uniquely determined by α, $\vartheta_\alpha(h) =: h_\alpha$ is, since h_α is the unique element of $\mathfrak{h}_\alpha$ such that

$$\overline{\alpha}(h_\alpha) = 2 . \tag{9}$$

There is a unique k-homomorphism $Sl_2 \to R$ compatible with ϑ_α, which induces an isogeny $\Theta_\alpha : Sl_2 \to S_\alpha$. This is seen e.g. by following ϑ_α by some faithful representation of R in view of the fact that every representation of $\mathfrak{sl}_2$ is compatible with a unique representation of Sl_2.

Let B_2 be the Borel subgroup of Sl_2 of upper triangular matrices of determinant one. For a representation space V of B define $V^{\beta+\mathbb{Z}\alpha} = \oplus V^{\gamma}$, $\gamma \in \beta + \mathbb{Z}\alpha, \alpha,\beta \in X^*(T)$. Conditions 1) and 2) of 6.6.2 hold for R and V iff they hold for Sl_2 and all the B_2-modules $V^{\beta+\mathbb{Z}\alpha}$ given by $\varphi \circ \Theta_\alpha | B_2$, $\beta \in X^*(T)$, $\alpha \in \Phi_+$.

So the necessity part of 6.6.2 follows from the special case $R = Sl_2$ proved above.

To prove sufficiency assume from now on that conditions 1) and 2) of 6.6.2 hold. Then the representation on $V^{\beta+\mathbb{Z}\alpha}$ given by $\varphi \circ \Theta_\alpha | B_2$ extends uniquely to a k-representation ρ of Sl_2 on $V^{\beta+\mathbb{Z}\alpha}$. Since the kernel of Θ_α is central, hence contained in B_2, ρ induces a k-representation φ_α of S_α on every $V^{\beta+\mathbb{Z}\alpha}$ and hence on V. So there is a unique k-representation φ_α of S_α on V extending $\varphi | S_\alpha \cap B$. The map φ_α is unique since $\varphi_\alpha \circ \Theta_\alpha$ is uniquely determined by its restriction to B_2.

At this point we see the uniqueness of the extended representation of R on V, since R is generated by B and the groups S_α, $\alpha \in \Phi_+$ ([22] 2.3). So it remains to show that the homomorphisms Φ_α on S_α and φ on B fit together to define a homomorphism from R to $Gl(V)$. To prove this we first look at the corresponding Lie algebra homomorphisms.

Analogous statements as above hold for the Lie algebras, see 6.6.4. There is a unique representation of $\mathfrak{s}_\alpha$ on V, namely $d\varphi_\alpha$, whose restriction to $\mathfrak{s}_\alpha \cap \mathfrak{h}$ is $d\varphi | \mathfrak{s}_\alpha \cap \mathfrak{h}$. So there is a unique linear map $\psi : \mathfrak{r} \to End(V)$, whose restriction to $\mathfrak{h}$ is $d\varphi$ and whose restriction to every $\mathfrak{s}_\alpha$ is a Lie algebra homomorphism. We shall prove that ψ is a Lie algebra homomorphism. ψ is obviously defined over k.

6.6.7 CLAIM *ψ is a Lie algebra homomorphism.*

For $\alpha \in \Phi_+$ let $\vartheta_\alpha : \mathfrak{sl}_2 \to \mathfrak{s}_\alpha$ be an isomorphism. Define

$X_\alpha = \vartheta_\alpha(x)$, $X_{-\alpha} = \vartheta_\alpha(y)$, $h_\alpha = \vartheta_\alpha(h)$. I show that for α,β in Δ the following equations hold

$$[\psi(h_\alpha),\psi(h_\beta)] = 0 \tag{10}$$

$$[\psi(h_\alpha),\psi(X_\beta)] = \overline{\beta}(h_\alpha)\psi(X_\beta) \tag{11}$$

$$[\psi(h_\alpha),\psi(X_{-\beta})] = -\overline{\beta}(h_\alpha)\psi(X_{-\beta}) \tag{12}$$

$$[\psi(X_\alpha),\psi(X_{-\alpha})] = \psi(h_\alpha) \tag{13}$$

$$[\psi(X_{-\alpha}),\psi(X_\beta)] = 0 \quad \text{for} \quad \alpha \neq \beta \tag{14}$$

$$(\mathrm{ad}\,\psi(X_\alpha))^{1-\overline{\beta}(h_\alpha)}\,\psi(X_\beta)\ 0 \quad \text{for} \quad \alpha \neq \beta \tag{15}$$

$$(\mathrm{ad}\,\psi(X_{-\alpha}))^{1-\overline{\beta}(h_\alpha)}\psi(X_{-\beta}) = 0 \quad \text{for} \quad \alpha \neq \beta \tag{16}.$$

Once these equations are shown, it follows from the Kac-Moody-type description of the derived Lie algebra of r ([24] VIII §4), due to Serre, that there is a unique Lie algebra homomorphism $\overline{\psi} : r \to \mathrm{End}(V)$ such that the restriction of ψ and $\overline{\psi}$ to $\bigoplus_{\alpha\in\Delta} r^\alpha \oplus \bigoplus_{\alpha\in\Delta} r^{-\alpha} \oplus \mathfrak{t}$ coincide, hence $\psi = \overline{\psi}$ and ψ is a Lie algebra homomorphism.

So it remains to prove the equations (10) - (16). The equations (10), (11) and (15) are true, since $\psi|s_\alpha$ is a Lie algebra homomorphism.

(12): Let h be an element of $\mathfrak{t}$ commuting with X_β , i.e. $\overline{\beta}(h) = 0$. Then $\Psi(h) : V \to V$ is a morphism of $s_\beta \cap h$-modules, hence of the s_β-module V into itself (see 6.6.5), so $\psi(h)$ commutes with $\psi(X_{-\beta})$, so

$$[\psi(h),\psi(X_{-\beta})] = -\overline{\beta}(h)\psi(X_{-\beta})$$

for $h \in \ker\overline{\beta}$. For $h = h_\beta$ this formula also holds, so it holds for every $h \in \mathfrak{t}$, by linearity, in particular for $h = h_\alpha$, $\alpha \in \Delta$.

(14): Look at the α-string through β : $St(\beta;\alpha) = \{\gamma \in \Phi ; \gamma \in \beta + \mathbb{Z}_\alpha\}$. Since every root is a linear combination of elements of Δ with only non-negative or with only non-positive coefficients, we have $\{\gamma \in \Phi ; \gamma \in \beta + \mathbb{Z}\alpha\} \subset \Phi \cap (\beta + \mathbb{N}_0\alpha) \subset \Phi_+$. So $\mathfrak{st}(\beta,\alpha) := \bigoplus_{\gamma\in St(\beta,\alpha)} r^\gamma \subset n_+$

and $\psi|\mathfrak{st}(\beta,\alpha) : \mathfrak{st}(\beta,\alpha) \to \mathrm{End}\, V$ is a $\mathfrak{h} \cap \mathfrak{s}_\alpha$ - module homomorphism between $\mathfrak{s}_\alpha$-modules, hence a homomorphism of $\mathfrak{s}_\alpha$-modules, in particular

$$[\psi(X_{-\alpha}),\psi(X_\beta)] = \psi[X_{-\alpha},X_\beta] = 0 .$$

(16): Let W be the $\mathfrak{s}_\alpha$ - submodule of $\mathrm{End}(V)$ generated by $\psi(X_{-\beta})$. We have $[\psi(X_\alpha),\psi(X_{-\beta})] = 0$ by (15) and $[\psi(h_\alpha),\psi(X_{-\beta})] = -\bar{\beta}(h_\alpha)\psi(X_{-\beta})$. So for the $\mathfrak{sl}_2$-module W defined by $\vartheta_\alpha : \mathfrak{sl}_2 \to \mathfrak{s}_\alpha$ we have $\psi(X_{-\beta}) \in W^{-\bar{\beta}(h_\alpha)} \subset \ker x$, hence $y^{1-\bar{\beta}(h_\alpha)}\psi(X_{-\beta}) = 0$, by 6.6.3, i.e. $(\mathrm{ad}\,\psi(X_{-\alpha}))^{1-\bar{\beta}(h_\alpha)}\psi(X_{-\beta}) = 0$.

This finishes the proof that ψ is a Lie algebra homomorphism.

6.6.8 End of the proof

Our claim is that there is a k-representation Ψ of R extending φ. We have already the unique k-representation $\psi : \mathfrak{r} \to \mathfrak{gl}(V)$ extending $d\varphi : \mathfrak{h} \to \mathfrak{gl}(V)$. We may assume that the $\mathfrak{r}$-module V is irreducible, because the k-projection to an irreducible submodule V' of V commutes with $\mathfrak{h}$, hence with B and with every $\mathfrak{s}_\alpha$, hence with every S_α (see [18] II § 7). So the B-module V' fulfills conditions 1) and 2) of 6.6.2 (see 6.6.6).

First assume R is semisimple. We want to find the graph of Ψ. The group $R \times GL(V)$ acts on its Lie algebra $\mathfrak{r} \times \mathfrak{gl}(V)$ by the adjoint representation. The graph $\mathfrak{g}$ of ψ is a k-subspace of $\mathfrak{r} \times \mathfrak{gl}(V)$. Let A be the k-subgroup of elements of $R \times GL(V)$ mapping $\mathfrak{g}$ to itself and inducing an automorphism of the Lie algebra $\mathfrak{g}$. Let π_i be the projection of $R \times GL(V)$ to its i-th component, $i = 1,2$. Then $\ker(\pi_1|A)$ is the group of homotheties of V, since ψ is irreducible. We show that the connected component C of $A \cap R \times SL(V)$ is the desired graph of Ψ. We know that $\varphi|B \cap S_\alpha$ extends to S_α, $\alpha \in \Phi_+$, and $\varphi(S_\alpha) \subset SL(V)$. Also $\varphi(B) \subset SL(V)$ since the element $\det \circ \varphi|T \in X^*(T)$ equals $\Sigma(\dim V^\alpha)\cdot\alpha$, $\alpha \in X^*(T)$, and is fixed by the

Weyl group by condition 2) of 6.6.2, consequently zero, since R is semisimple. So $\pi_1(C)$ contains B and S_α, $\alpha \in \phi_+$, hence $\pi_1(C) = R$ ([22] 2.3). Now $\pi_1|C : C \to R$ is an isogeny, since $\ker \pi_1 \cap C$ is a group of homotheties of V of determinant one, and has a section $\varphi|T$ over the maximal torus T of R, hence is an isomorphism. Now define $\Psi = \pi_2 \circ (\pi_1|C)^{-1}$.

If R is not semisimple, first extend $\varphi|R' \cap B$ to $\bar{\varphi} : R' \to GL(V)$. For every element t of the center $Z(R) \subset B$ of R ([18] 11.11) of R, $\varphi(t)$ commutes with $\varphi|R' \cap B$ hence with $\bar{\varphi}$, by uniqueness of the extension. This implies that for $t \in Z(R)$, $x \in R'$, the formula $\bar{\varphi}(t \cdot x) = \varphi(t) \cdot \bar{\varphi}(x)$ gives a well-defined k-representation of R □

In 6.5 the following corollary of theorem 6.6.2 was used.

6.6.9 COROLLARY *Suppose a* k-*representation of* B *on* V *and a* k-*representation of* R *on* W *are given. Let* $\pi : V \to W$ *be a surjective homomorphism of* B-*modules, whose kernel is contained in* V^0, *the subspace of* T-*fixed vectors. Then the representation of* B *on* V *extends to a unique* (k-)*representation of* R *on* V. *The map* π *is a morphism of* R-*modules.* R *acts trivially on* $\ker \pi$ *and there is a complementary* R-*submodule of* $\ker \pi$ *in* V.

PROOF. The first claim follows from 6.6.2, since our conditions are trivial for $\beta = 0$ and hold for $\beta \neq 0$, since $V^\beta \tilde{\to} W^\beta$ and π is a morphism of n_+-modules. The second claim follows from the fact that a B-module homomorphism between R-modules is actually a R-module homomorphism, cf. 6.6.5. So $\ker \pi$ is a R-submodule whose only weight is zero, hence a trivial R-module. The last claim holds, since every k-representation of R is completely reducible □

VII. S-ARITHMETIC SOLVABLE GROUPS

In this capter we prove the following theorems. Let k be a finite extension field of $\mathbb{Q}$, let G be a connected solvable k-group.

7.0.1 THEOREM *Let* Γ *be a S-arithmetic subgroup of* G. *Then* Γ *has a finite presentation iff the following two conditions hold.*

1) Γ *is tame.*

2) $H_2(\Gamma';\mathbb{Z})$ *is a finitely generated* $\mathbb{Z}\Gamma$*-module.*

7.0.2 THEOREM *Let* Γ *be a S-arithmetic subgroup of* G. *Suppose* G *is k-split. Then* Γ *has a finite presentation iff the following two conditions hold.*

1) Γ *is tame.*

2) $H_2(\Gamma;\mathbb{Z})$ *is a finitely generated abelian group.*

The definition of tameness will be recalled below (see 7.3.10).

In case G is not connected, we have the following result.

7.5.2 THEOREM *Let* Γ *be a S-arithmetic subgroup of a solvable k-group,* Δ *a normal nilpotent subgroup of* Γ *such that* Γ/Δ *contains a finitely generated abelian subgroup* Q *of finite index. Then* Γ *has a finite presentation iff the following two conditions hold.*

1) Δ^{ab} *is a tame* $\mathbb{Z}Q$*-module.*

2) $H_2(\Delta;\mathbb{Z})$ *is a finitely generated* $\mathbb{Z}\Gamma$*-module.*

REMARK 1 Note that conditions 1) and 2) of either theorem are purely group-theoretic and make no reference to the algebraic group structure. This may be considered an advantage. Though, I believe that in practice the linearized versions of conditions 1) and 2) as given in theorem 6.2.3 making use of the adjoint action of G on its unipotent radical are easier to check.

REMARK 2 7.5.2 is more general than 7.0.1 and may have advantages in applications even for connected G because it says that one does not have to take $\Delta = G'$.

REMARK 3 The proofs actually show that a group Γ as in either theorem is *finitely presented iff it is of type* FP_2 (7.3.10, 7.4.10).

REMARK 4 H. Åberg has stated 7.5.2 with a sketch proof in [8]. I announced 7.0.2 at Oberwolfach in 1986 and was later on able to get hold of a copy of Åberg's preprint.

The proof of the theorems consists in deriving them from the results of 6.2. So once again we have to compare $H_*(\Gamma)$ and $H_*(\mathfrak{u})$.

The contents of the present chapter are as follows. 7.1 contains facts about S-arithmetic groups we shall need later on. In 7.2 we describe the commutator subgroup of a S-arithmetic subgroup of a connected solvable k-group. As a corollary we obtain.

7.2.9 THEOREM *Let* Γ *be a S-arithmetic subgroup of a connected solvable k-group* G . *Then* Γ *is finitely generated iff* Γ^{ab} *is finitely generated and the natural* $\mathbb{Z}\Gamma$ *-module* Γ'/Γ'' *is finitely generated.*

In 7.3 we suppose that a k-representation of a k-torus T on a vector space M defined over k is given and compare compact generation of the T_{k_p}-modules M_{k_p} , $p \in S$, and finite generation in the corresponding S-arithmetic situation and also compare the corresponding Bieri-Strebel invariants. This gives the equivalence of condition 1) in

7.0.1 or 7.0.2 and condition 1) in 6.2.3. In 7.4 we compare $H_2(\Gamma)$ and $H_2(\mathfrak{u}_k)$. This gives sufficiency of condition 2) in 7.0.1 or 7.0.2 for condition 2) of 6.2.3. All this together gives a proof of theorems 7.0.1 and 7.0.2 above (see 7.4.8). In 7.5 we deal with the case that G is not connected and prove theorem 7.5.2.

7.1 FACTS ABOUT S-ARITHMETIC GROUPS

In this section we fix the notations for the whole chapter and recall some known facts about S-arithmetic groups.

7.1.1 Let k be an algebraic number field of finite degree over $\mathbb{Q}$, $\mathcal{O}$ the ring of integers of k, V the set of primes (or equivalence classes of absolute values) of k, $P \subset V$ the set of finite primes, $x \to \|x\|_v$ the normalized absolute value associated to $v \in V$, k_v the completion of k with respect to $\|\cdot\|_v$, $\mathcal{O}_p$ $(p \in P)$ the ring of p-adic integers of k_p and A_k or A the ring of adeles of k. If $v \in V-P$ is an infinite (= archimedean) prime, we put $\mathcal{O}_v = k_v$. If S is a subset of V, then $\mathcal{O}(S) = \{x \in k \mid x \in \mathcal{O}_v \text{ for } v \in V-S\}$.

7.1.2 The letter G will always denote a linear algebraic group over k. Usually we think of G as embedded into a fixed general linear group GL_n. Given a subring B of a field containing k, G_B is the group of elements of G with entries from B and determinant invertible in B. We shall write G_v for G_{k_v}. For a finite subset S of V put $G_S = \prod_{v \in S} G_v$. We write $G_\infty := G_{V-P}$ and $G_{S\cup\infty} := G_{S\cup(V-P)}$. Let j_S be the diagonal embedding $G_k \to G_S$, $S \neq \emptyset$.

A subgroup Γ of G_k is called S-*arithmetic* if Γ is commensurable with $G_{\mathcal{O}(S)}$, i.e. $\Gamma \cap G(\mathcal{O}(S))$ is of finite index in both Γ and $G_{\mathcal{O}(S)}$. If $S \subset V - P$ a S-arithmetic group is called *arithmetic*.

7.1.3 PROPOSITION G_S *modulo the closure of* $j_S(\Gamma)\cdot G_\infty$ *is compact for every* S-*arithmetic group* Γ , *if* $S \supset V - P$.

PROOF. This follows immediately from the basic result of Borel's [16] Theorem 5.1 that the set of distinct double cosets $G_A^\infty \backslash G_A / G_k$ is finite, where $G_A^\infty = G_\infty \times \prod_{p \in P} G_{o_p}$ and G_k is diagonally embedded into G_A . In fact, put $G_{A(S)} = G_S \times \prod_{p \notin S} G_{o_p}$, then $G_A^\infty \backslash G_{A(S)} / (G_k \cap G_{A(S)})$ is finite, but $G_k \cap G_{A(S)} = G_{o(S)}$ is commensurable with Γ and $x^{-1} \cdot G_A^\infty \cdot x$ is contained in a finite number of cosets of G_A^∞ for every $x \in G_A$, hence $G_{A(S)} = G_A^\infty C \Gamma = D G_A^\infty \Gamma$ for finite subsets C , D of $G_{A(S)}$. Projecting G_A to G_S gives the desired result □

Project onto G_p to obtain

7.1.4 COROLLARY $G_p / \overline{j_p(\Gamma)}$ *is compact for every* S-*arithmetic subgroup* Γ *of* G *if* $p \in P \cap S$ □

Similarly as in 7.1.3, the elementary case of a unipotent group of [16] Theorem 5.8 gives

7.1.5 *If* G *is a unipotent* k-*group and* Γ *is a* S-*arithmetic subgroup, then* $j_{S \cup \infty} : \Gamma \to G_{S \cup \infty}$ *has discrete cocompact image.*

7.1.6 PROPOSITION *Let* $f : G \to G'$ *be a surjective* k-*morphism of linear algebraic groups. Let* Γ *be a* S-*arithmetic subgroup of* G . *Then* $f(\Gamma)$ *is a* S-*arithmetic subgroup of* G' .

In Borel [17] Theorem 6 a proof is given for arithmetic groups. An analogous proof for S-arithmetic groups can be given using the special case of an isogeny proved in [16] Proposition 8.12 □

7.1.7 *Let* $p \in V$. *A* k-*torus* T *contains exactly one maximal* k_p-*split subtorus* D *and* T_{k_p} / D_{k_p} *is compact.*

This is a special case of [22] Theorem 8.2 □

7.2 THE COMMUTATOR SUBGROUP

In this section we describe the commutator subgroup of a S-arithmetic subgroup of a connected solvable k-group. As a corollary we obtain a criterion for Γ to be finitely generated.

7.2.1 It is not true that the commutator subgroup of an arithmetic subgroup of a connected k-group G is an arithmetic subgroup of G', e.g. the free group Γ on two generators is an arithmetic subgroup of SL_2 over $\mathbb{Q}$, but its commutator subgroup is of infinite index in Γ. Even for solvable G the commutator subgroups of two arithmetic subgroups Γ_1, Γ_2 may not be commensurable. E.g. let G be the semidirect product of SO_2 with its natural action on $U = \mathbb{G}_a \oplus \mathbb{G}_a$. Then $\Gamma_1 = R \oplus R \subset U$ and the product Γ_2 of Γ_1 with the subgroup $\{\pm 1\}$ of SO_2 are both arithmetic subgroups of G for $R = \mathbb{Z}$, but Γ_1 is abelian and Γ_2/Γ_2' is finite. For $R = \mathbb{Z}[\frac{1}{p}]$ we get an example of two commensurable S-arithmetic groups Γ_1, Γ_2 such that $H_1(\Gamma_2)$ is finite, but $H_1(\Gamma_1)$ is not finitely generated, if $S = \{p\}$ and SO_2 does not split over $\mathbb{Q}_p$.

7.2.2 Let G be a solvable k-group, let U be its unipotent radical. Let Γ be a S-arithmetic subgroup of G. Let ρ be the representation of G on the abelianized Lie algebra $\mathfrak{u}^{ab}$ of U induced by the adjoint representation.

7.2.3 PROPOSITION *The commutator subgroup* Γ' *of* Γ *is a S-arithmetic subgroup of the k-group* $(\Gamma U)'$.

We first single out the special case $G = U$.

7.2.4 PROPOSITION *If* Γ *is a S-arithmetic subgroup of the unipotent k-group* U, *then* Γ' *is a S-arithmetic subgroup of* U'.

PROOF. The claim is that Γ' is of finite index in the S-arithmetic subgroup $\Gamma \cap U'$ of U'. Let $U = U_0 \supset U_1 \supset \ldots \supset U_r \supset 0$ be the descending central series of U. The commutator induces a surjective k-morphism of abelian unipotent k-groups $U/U_1 \otimes U_i/U_{i+1} \to U_{i+1}/U_{i+2}$. Embed Γ into $U_{S\cup\infty}$ by $j_{S\cup\infty}$. Since $U_{i,S\cup\infty}/(\Gamma\cap U_i)$ is compact (7.1.5), it follows that $U_{i+1,S\cup\infty}/(\Gamma'\cap U_{i+1,S\cup\infty})\cdot U_{i+2,S\cup\infty}$ is compact and hence by descending induction that $U_{1,S\cup\infty}/\Gamma'$ is compact which proves our claim □

PROOF OF 7.2.3. We claim that $(\Gamma U)'$ is a k-group and that Γ' is of finite index in $\Gamma \cap (\Gamma U)'$. By the case $G = U$ settled already we may assume that $U' = 0$, i.e. that U is abelian. We identify U and its Lie algebra $\mathfrak{u}$ by the exponential map. The representation ρ induces a representation of the k-torus $T = G/U$ on U. We have

(7.2.5) $$M := (\Gamma U)' = \operatorname{span}\{(\rho(\gamma) - \mathrm{Id})U \mid \gamma\in\Gamma\},$$

hence M is a k-group.

I claim that

(7.2.6) $$M \oplus U^0 = U,$$

where $U^0 = \{u\in U \mid \rho(\gamma)u=u\}$. Let K be a T-submodule of U complementary to the T-module M. Then $\rho(\gamma) - \mathrm{Id}, \gamma\in\Gamma$, maps M and K to itself resp. and maps $U = K + M$ to M by 7.2.5, hence $K \subset U^0$. It follows that

(7.2.7) $$(\rho(\gamma)-\mathrm{Id})U = (\rho(\gamma)-\mathrm{Id})M .$$

If on the other hand N is a T-submodule of M complementary to $M\cap U^0$, then $(\rho(\gamma)-\mathrm{Id}))M \subset N$, hence $M = \operatorname{span}((\rho(\gamma)-\mathrm{Id})M \mid \gamma\in\Gamma\} \subset N$, so $M = N$ and $M \cap U^0 = 0$. This proves that $M \oplus U^0 = U$.

It remains to show that Γ' is of finite index in $\Gamma \cap M$. Let Δ be an $\mathcal{O}(S)$-module contained and of finite index in $\Gamma \cap M$. Then the group Δ_1 generated by $\{(\rho(\gamma)-\mathrm{Id})\delta \mid \gamma\in\Gamma, \delta\in\Delta\}$ is an $\mathcal{O}(S)$-module, is contained in $\Gamma \cap M$ and spans M by 7.2.7 since Δ spans M, hence Δ_1 is a S-arithmetic subgroup of M. So Γ' is S-arithmetic in M, since $\Delta_1 \subset \Gamma' \subset \Gamma \cap M$ □

7.2.8 REMARK Note the following corollary of the proof: *$(\Gamma U)'/U'$ and $K=\{u\in U^{ab} \mid \gamma u \gamma^{-1}=u$ for every $\gamma\in\Gamma\}$ are complementary subspaces of U^{ab}, hence $G/(\Gamma U)'$ is the semidirect product $G/U \ltimes K$* □

As a corollary we obtain

7.2.9 THEOREM *Let Γ be a S-arithmetic subgroup of a connected solvable k-group G. Then Γ is finitely generated iff Γ^{ab} is finitely generated and the $\mathbb{Z}\Gamma$-module Γ'/Γ'' is finitely generated.*

PROOF. If Γ is finitely generated, then so is the abelian group $\Gamma^{ab} = \Gamma/\Gamma'$, hence Γ^{ab} is finitely presented. So in the exact sequence

$$\Gamma' \rightarrowtail \Gamma \twoheadrightarrow \Gamma^{ab}$$

the kernel Γ' is finitely generated as a normal subgroup. In particular the $\mathbb{Z}\Gamma$-module Γ'/Γ'' defined by passing to quotients starting from inner automorphisms is finitely generated.

Conversely, $\Gamma' \subset U$ is nilpotent, so a subset of Γ' generates the group Γ' iff its image in Γ'^{ab} generates Γ'^{ab} (see 2.3.8). This implies the claim □

Another corollary is the following one.

7.2.10 COROLLARY *Suppose $S \neq \emptyset$. Suppose Γ is a S-arithmetic subgroup of the connected solvable k-group G. If Γ^{ab} is finitely generated, then Γ' is a S-arithmetic subgroup of the unipotent radical U of G.*

PROOF. Γ' is S-arithmetic in $(\Gamma U)'$, so $\Gamma/\Gamma' \to G/(\Gamma U)'$ has finite kernel and its image Δ is a S-arithmetic subgroup of $G/U \ltimes K$, by 7.1.6 and 7.2.8. On the other hand, if Γ^{ab} is finitely generated, hence so is its image Δ and hence so is the subgroup $\Delta \cap K$, a S-arithmetic subgroup of the vector space K defined over k. If $S \neq \emptyset$ this can only happen if $K = 0$. So $(\Gamma U)' = U$ by 7.2.8, so Γ' is S-arithmetic in U, by 7.2.3 □

7.3 The Bieri-Strebel invariant

In this section we describe the Bieri-Strebel invariant of certain actions. More presisely if a k-representation ρ of a k-torus T on a vector space M defined over k is given we compute the Bieri-Strebel invariant of the action of a S-arithmetic subgroup Θ of T on a S-arithmetic subgroup Λ of M in terms of the weights of the localized representations. As a corollary we obtain the equivalence of the conditions "Γ tame" and condition 1) of 6.2.3 (see 7.3.13).

7.3.1 Recall the following definition of the Bieri-Strebel invariant. We generalize it slightly to topological groups. Let Q be an abelian topological group. Let $\mathrm{Hom}(Q;\mathbb{R})$ be the $\mathbb{R}$-vector space of continuous homomorphisms $Q \to \mathbb{R}$. Two non-zero elements χ_1, χ_2 of $\mathrm{Hom}(Q;\mathbb{R})$ are called *equivalent* if there is a positive real number λ such that $\chi_2 = \lambda \cdot \chi_1$. The set of equivalence classes $[\chi]$ of non-zero elements $\chi \in \mathrm{Hom}(Q,\mathbb{R})$ is denoted $S(Q)$, because if Q is discrete and finitely generated then $S(Q)$ is homeomorphic to a sphere of dimension $\mathrm{rank}\, Q - 1$. For $\chi \in \mathrm{Hom}(Q,\mathbb{R})$ define the monoid

$$Q_\chi := \{q \in Q \mid \chi(q) \geq 0\} .$$

Now suppose a topological Q-module A is given, i.e. an abelian topological group A together with a continuous action of Q on A by automorphisms. The *Bieri-Strebel invariant* is by definition

$$\Sigma_A := \Sigma_A(Q) :=$$

$$\{[\chi] \in S(Q) \mid A \text{ is a compactly generated } \mathbb{Z}Q_\chi\text{-module}\} .$$

7.3.2 Let T be a k-torus, let M be a vector space defined over k, i.e. an algebraic group isomorphic to a direct sum of $\mathbb{G}_a$'s and defined over k. It is then k-isomorphic to a direct sum of $\mathbb{G}_a$'s. Let ρ be a k-representation of T on M. For every character $\chi \in X(T)$ the corresponding *weight space* V^χ is

$$M^\chi = \{m \in M \mid \rho(t)m = \chi(t)m \text{ for every } t \in T\} .$$

The set of $\chi \in X(T)$ such that $M^{\chi} \neq 0$ is called the *set of weights* of the representation ρ or of the T-module M and is denoted $W(\rho) = W(T,M)$. We have

$$M = \bigoplus_{\chi \in W(T,M)} M^{\chi} .$$

Let S be a finite subset of $V - P$. Suppose Θ is a S-arithmetic subgroup of T and Λ is a S-arithmetic subgroup of M such that Λ is Θ-invariant, i.e. $\rho(\vartheta)\Lambda \subset \Lambda$ for every $\vartheta \in \Theta$.

The following lemma reduces our question to the individual primes in S.

7.3.3 LEMMA *a)* Λ *is a finitely generated* $\mathbb{Z}\Theta$*-module iff* M_S *is a compactly generated* $\mathbb{Z}\Theta$*-module iff* M_p *is a compactly generated* $\mathbb{Z}\Theta$*-module for every* $p \in S$.

b)
$$\Sigma_{\Lambda}(\Theta) = \Sigma_{M_S}(\Theta) = \bigcap_{p \in S} \Sigma_{M_p}(\Theta) .$$

PROOF. M_S/Λ is compact (7.1.4). This implies the first claims of a) and b). The second ones follow from the following facts. M_∞ is compactly generated as a topological group, even without operators. For $B = \bigoplus_{i=1}^{m} A_i$ we have

(7.3.4)
$$\Sigma_B(Q) = \bigcap_{i=1}^{m} \Sigma_{A_i}(Q) \;\square$$

(7.3.5) Now fix $p \in S$. Let $j_p : \Theta \to T_p$ be the inclusion and let Θ_0 be the kernel of the map $\Theta \to T_p \to T_p / T_{0_p}$. For a subgroup Q_0 of an abelian topological group Q put

$$\mathrm{Hom}(Q,Q_0;\mathbb{R}) = \{\chi \in \mathrm{Hom}(Q;\mathbb{R})\,;\,\chi|Q_0 = 0\}$$

and let $S(Q,Q_0) \subset S(Q)$ be the corresponding set of equivalence classes of non-zero elements $\chi \in \mathrm{Hom}(Q,Q_0;\mathbb{R})$. Let D be the maximal k_p-split subtorus of T and let $i : D \to T$ be the inclusion. Every $\chi \in X(D)$ is defined over k_p, since D is k_p-split, hence induces a map $\chi_p : D_p \to GL_{1,p} = k_p^*$. Composing with the natural valuation

$v_p : k_p^* \to \mathbb{R}$ gives a map $v_p \circ \chi_p \in \mathrm{Hom}(D_p;\mathbb{R})$. We thus get a map of vector spaces over $\mathbb{R}$

$$X(D) \otimes_{\mathbb{Z}} \mathbb{R} \longrightarrow \mathrm{Hom}(D_p;\mathbb{R})$$
$$\chi \otimes r \longmapsto r\cdot(v_p \circ \chi_p) \quad .$$

7.3.6 PROPOSITION *M_p is a compactly generated $\mathbb{Z}\Theta$-module iff 0 is not a weight of D on M iff zero is the only element of M_p having a relatively compact T_p-orbit.*

7.3.7 PROPOSITION *For fixed $p \in S$ we have isomorphisms:*

$$X(D) \otimes_{\mathbb{Z}} \mathbb{R} \to \mathrm{Hom}(D_p;\mathbb{R}) \xleftarrow{i^*} \mathrm{Hom}(T_p;\mathbb{R}) \xrightarrow{j^*} \mathrm{Hom}(\Theta,\Theta_o;\mathbb{R}) \ .$$

For the induced maps of equivalence classes we get a bijection

$$\text{complement of } \{[\chi\otimes 1];\chi\in W(D,M)\} \simeq \Sigma_{M_p}(\Theta) \cap S(\Theta,\Theta_o) \ .$$

If furthermore M_p is a compactly generated $\mathbb{Z}\Theta$-module then

$$S(\Theta) \setminus S(\Theta,\Theta_o) \subset \Sigma_{M_p}(\Theta) \ .$$

In view of 7.1.4 and 7.1.7 these propositions are implied by the next two lemmas.

7.3.8 LEMMA *Let Q be an abelian topological group, let U be an open compact subgroup of Q and let R be a subgroup of Q such that $Q/\bar{R}$ is compact. Then the inclusion $i : R \to Q$ induces an isomorphism*

$$i^* : \mathrm{Hom}(Q;\mathbb{R}) \to \mathrm{Hom}(R,R\cap U;\mathbb{R}) \quad .$$

b) For every topological Q-module A, i^* induces a bijection

$$\Sigma_A(Q) \simeq \Sigma_A(R) \cap S(R,R\cap U) \ .$$

c) A locally compact topological Q-module A is compactly generated as a Q-module iff A is compactly generated as a R-module. In this case

$$S(R) \setminus S(R,R\cap U) \subset \Sigma_A(R) \ .$$

PROOF. It is easy to prove a): Every $\chi \in \mathrm{Hom}(R,R\cap U;\mathbb{R})$ first extends uniquely to $R\cdot U$, then to the supergroup Q of finite index. And every $\chi \in \mathrm{Hom}(Q;\mathbb{R})$ vanishes on U. Concerning b), $i^*\Sigma_A(Q) \supset \Sigma_A(R) \cap S(R,R\cap U)$ is trivial. Conversely, if $\chi \in \mathrm{Hom}(Q;\mathbb{R})$ then $Q_\chi \subset \bigcup_j (RU)_\chi q_j$ where $\{q_j\}$ is a finite set of representatives for Q/RU such that $\chi(q_j) \leq 0$. Hence if A is compactly generated as $\mathbb{Z}Q_\chi$-module, it is so as $\mathbb{Z}(RU)_\chi$-module and hence as $\mathbb{Z}R_\chi$-module, since $(RU)_\chi = R_\chi \cdot U$ with compact U. The first claim of c) is obvious and the last one follows from the fact that $RU = R_\chi \cdot U$ is of finite index in Q if $\chi | R\cap U \neq 0$ □

7.3.9 Lemma *Let* M *be the one dimensional representation of* D *corresponding to* $\chi \in X(D)$ *. It is defined over* k_p *. The* $\mathbb{Z}D_p$*-module* M_p *is compactly generated iff* $\chi \neq 0$ *. In this case*

$$\Sigma_{M_p}(D_p) \simeq \text{complement of } \{[\chi \otimes\ ]\} \text{ in } X(D) \otimes_{\mathbb{Z}} \mathbb{R} .$$

Proof. If $\chi \neq 0$ then the image $\chi_p(D_p)$ contains an element $\chi_p(d)$ with $v_p(\chi_p(d)) < 0$ which thus is expanding (1.2.1) on M_p, so M_p is a compactly generated $\mathbb{Z}D_p$-module. For $\lambda = r\cdot(v_p \circ \chi)$, $r > 0$, the image of $D_{p,\lambda} = \{d \in D_p \mid r(v_p\circ\chi)(d) \geq 0\}$ under $\chi_p : D_p \to GL_{1,p} = k_p^*$ is contained in the compact group $\mathcal{O}_p^*$, so M_p is not compactly generated as a $D_{p,\lambda}$-module. Conversely, if $\lambda \in \mathrm{Hom}(D_p;\mathbb{R}) \setminus \{0\}$ is not a positive multiple of $v_p \circ \chi$, then there is an element $d \in D_p$ such that $v_p \circ \chi(d) < 0$ and $\lambda(d) > 0$ (cf. the proof of 3.4.2), hence k_p is compactly generated as a $D_{p,\lambda}$-module, since $\bigcup_{n\in\mathbb{N}} \chi(d^n)\mathcal{O}_p = k_p$ □

7.3.10 The topological Q-module A is called *tame* if

$$\Sigma_A(Q) \cup -\Sigma_A(Q) = S(Q) ,$$

where $-\Sigma = \{[-\chi];[\chi]\in\Sigma\}$ for $\Sigma \subset S(Q)$. A group Γ will be called *tame* if Γ is a finitely generated group and the $\mathbb{Z}\Gamma^{ab}$-module Γ'/Γ'' is tame. It is a basic fact, proved by Bieri and Strebel [14], that every finitely presented solvable group Γ is tame. We shall need this fact only for S-arithmetic subgroups of connected solvable k-groups. A proof for this case follows from 6.2.3 and 7.3.12. Actually, Bieri and Strebel proved that a solvable group of type FP_2 is tame.

7.3.11 Let G be a connected solvable k-group, U the unipotent radical of G, Γ a S-arithmetic subgroup of G. For every $p \in S$ let H be the maximal k_p-split subgroup of G and $D = H/U$ the maximal k_p-split subtorus of $T = G/U$. Let ρ be the representation of T on $\mathfrak{u}^{ab}$ induced by the adjoint representation. Let $W \subset X(D)$ be the set of weights of $\rho : D \to GL(\mathfrak{u}^{ab})$. Recall 6.2.5:

7.3.12 *Γ is finitely generated iff, for every $p \in S$, zero is not in W* □

7.3.13 PROPOSITION *Γ is tame iff, for every $p \in S$, any two (not necessarily distinct) elements of W are positively independent in $X(D) \otimes_{\mathbb{Z}} \mathbb{R}$ (condition 1 of theorem 6.2.3).*

PROOF. Another way of stating condition 1 is: For every $p \in S$ we have $0 \notin W$ and $[W] \cap -[W] = \emptyset$ in $X(D) \otimes_{\mathbb{Z}} \mathbb{R} \smallsetminus \{0\}$ modulo equivalence.

Both conditions of 7.3.13 imply that Γ is finitely generated. So $\Gamma^{ab} \to T$ and $\Gamma'/\Gamma'' \to \mathfrak{u}^{ab}$ have finite kernel and S-arithmetic image Θ and Λ, say (7.2.10, 7.2.4, 7.1.6). In particular, the $\mathbb{Z}\Gamma^{ab}$-module Γ'/Γ'' is tame iff the $\mathbb{Z}\Theta$-module Λ is tame. So necessity of condition 1 is implied by 7.3.3 and 7.3.7. To prove the converse, let us denote the kernel of $\Theta \to T_p \to T_p/T_{o_p}$ by $\Theta_{o,p}$ for $p \in S$. If $[\chi] \notin S(\Theta,\Theta_{o,p})$ for every $p \in S$, then $[\chi] \in \Sigma_\Lambda(\Theta)$ by 7.3.3 and 7.3.6. If $[\chi] \notin S(\Theta,\Theta_{o,p})$ for every $p \in S$ except one, then $[\chi] \in \Sigma_\Lambda(\Theta) \cup -\Sigma_\Lambda(\Theta)$ by 7.3.3, 7.3.6 and condition 1. So it remains to prove that every $\chi \neq 0$ in $\mathrm{Hom}(\Theta,\mathbb{R})$ vanishes on $\Theta_{o,p}$ for at most one $p \in S$. If $0 \neq \chi \in \mathrm{Hom}(\Theta,\Theta_{o,p};\mathbb{R})$, then χ extends uniquely to a non-zero $\chi_p \in \mathrm{Hom}(T_p;\mathbb{R})$ by 7.3.8 and 7.1.4. Put $S_1 = S - \{p\}$. Then χ vanishes on the S_1-arithmetic subgroup Θ_o, hence the only continuous extension of χ to $T_{p'}$, $p' \in S_1$, is zero by 7.3.8 and 7.1.4. So there is at most one p such that our non zero χ vanishes on $\Theta_{o,p}$ □

7.4 The second homology

In this section we show that for a S-arithmetic subgroup Ω of a unipotent k-group U with Lie algebra $\mathfrak{u}$ the image of $H_i(\Omega;\mathbb{Z})$ in $H_i(\mathfrak{u}_k;k)$ is S-arithmetic for arbitrary $i > 0$ if $k = \mathbb{Q}$ and for $0 < i \leq 2$ if $[k:\mathbb{Q}]$ is finite. This will imply the remaining parts of the proof of 7.0.1 and 7.0.2.

The case $k = \mathbb{Q}$ is much easier and gives a more general result. So we give a separate proof for this case.

7.4.1 Proposition *Let* $k = \mathbb{Q}$. *For a S-arithmetic subgroup* Ω *of a unipotent* $\mathbb{Q}$*-group* U *with Lie algebra* $\mathfrak{u}$, *the map*

$$H_i(\Omega;A) \to (\Omega;A \otimes_{\mathbb{Z}} \mathbb{Z}(S))$$

induced by the coefficient homomorphism is an isomorphism for $i > 0$ *and any trivial* $\mathbb{Z}\Gamma$*-module* A . *Furthermore* $H_*(\Omega;\mathbb{Z}(S))$ *is a finitely generated* $\mathbb{Z}(S)$*-module.*

Proof. Note that a $\mathbb{Z}(S)$-module is a uniquely p-divisible group for every $p \in S$. Conversely, any uniquely p-divisible abelian group for every $p \in S$ can be turned into a $\mathbb{Z}(S)$-module in a unique way. If U is abelian, then $H_i(\Omega;A) = \Lambda^i_{\mathbb{Z}}\Omega \otimes A = \Lambda^i\Omega \otimes A \otimes \mathbb{Z}(S) = H_i(\Omega;A\otimes\mathbb{Z}(S))$, since Ω is a $\mathbb{Z}(S)$-module in a unique way. For arbitrary U let Z be the center of U and use the spectral sequences

$$H_i(\Omega/\Omega\cap Z;H_j(\Omega\cap Z;B)) \Rightarrow H_{i+j}(\Omega;B)$$

for $B = A$ and $B = A \otimes \mathbb{Z}(S)$ resp. to obtain the result. The second claim is clear for abelian U and follows by induction from the spectral sequence above for arbitrary U , since $\mathbb{Z}(S)$ is noetherian □

7.4.2 Corollary *The image of* $H_i(\Omega;\mathbb{Z})$ *in* $H_i(\Omega;\mathbb{Q}) \simeq H_i(\mathfrak{u}_{\mathbb{Q}};\mathbb{Q})$ *is* S*-arithmetic for* $i > 0$.

Proof. The image is a finitely generated $\mathbb{Z}(S)$-module which spans

$H_i(\Omega;\mathbb{Q})$ and $H_i(\Omega;\mathbb{Q}) \approx H_i(\mathfrak{u}_{\mathbb{Q}};\mathbb{Q})$ by [41, 51 Theorem 8.1] □

Now let k be a finite extension field of $\mathbb{Q}$.

7.4.3 PROPOSITION *Suppose Ω is a S-arithmetic subgroup of the unipotent k-group U . Then the image of the natural map $\sigma : H_2(\Omega;\mathbb{Z}) \to H_2(\mathfrak{u}_k;k)$ is a S-arithmetic subgroup of $H_2(\mathfrak{u}_k;k)$.*

PROOF. Recall the definition of σ (5.3.6). Let F be the free group with basis Ω , R the kernel of the canonical map $F \to \Omega$,

$$R \rightarrowtail F \twoheadrightarrow \Omega \ .$$

This sequence induces exact sequences

$$E \ : \quad R/(F,R) \rightarrowtail F/(F,R) \twoheadrightarrow \Omega$$

and

$$E' \ : \ (F'\cap R)/(F,R) \rightarrowtail F'/(F,R) \twoheadrightarrow \Omega' \ .$$

We have called E' the Hopf sequence. $(F'\cap R)/(F,R)$ is naturally isomorphic to $H_2(\Omega;\mathbb{Z})$.

Analogously let $\mathfrak{f}$ be a Lie algebra over k together with a k-linear map $\mathfrak{u}_k \to \mathfrak{f}$, such that every k-linear map of $\mathfrak{u}_k$ to any Lie algebra $\mathfrak{g}$ over k gives rise to a unique commutative triangle

$$\begin{array}{ccc} \mathfrak{u}_k & \longrightarrow & \mathfrak{f} \\ & \searrow & \downarrow \\ & & \mathfrak{g} \end{array}$$

with a unique homomorphism $\mathfrak{f} \to \mathfrak{g}$ of Lie algebras over k . So $\mathfrak{f}$ is the free Lie algebra over k with any k-basis of the k-vector space $\mathfrak{u}_k$ as set of free generators. Let $\mathfrak{r}$ be the kernel of the map $\mathfrak{f} \to \mathfrak{u}_k$ induced by the identity $\mathfrak{u}_k \to \mathfrak{u}_k$,

$$\mathfrak{r} \rightarrowtail \mathfrak{f} \twoheadrightarrow \mathfrak{u}_k \ .$$

Again, it induces exact sequences of Lie algebras over k

$$\mathfrak{E} \ : \ \mathfrak{r}/[\mathfrak{f},\mathfrak{r}] \rightarrowtail \mathfrak{f}/[\mathfrak{f},\mathfrak{r}] \twoheadrightarrow \mathfrak{u}_k$$

and

$\mathfrak{E}' : (\mathfrak{f}'\cap\mathfrak{r})/[\mathfrak{f},\mathfrak{r}] \rightarrowtail \mathfrak{f}'/[\mathfrak{f},\mathfrak{r}] \twoheadrightarrow \mathfrak{u}_k'$.

We have called $\mathfrak{E}'$ the Hopf sequence of $\mathfrak{u}_k$. The vector space $(\mathfrak{f}'\cap\mathfrak{r})/[\mathfrak{f},\mathfrak{r}]$ is naturally isomorphic to $H_2(\mathfrak{u}_k;k)$.

Put $\mathfrak{g} = \mathfrak{f}/[\mathfrak{f},\mathfrak{r}]$. Note that $\mathfrak{E}$ is a central extension, so $\mathfrak{g}$ is nilpotent, hence $\exp \mathfrak{g}$ is a well defined Lie group over k (2.5) and also $\exp \mathfrak{g}$ is the group of k-points G_k of a unipotent k-group G , unique up to k-isomorphism (cf. [26]. The map

$$\Omega \xrightarrow{\log} \mathfrak{u}_k \to \mathfrak{f} \to \mathfrak{g} \xrightarrow{\exp} G_k$$

extends to a unique group homomorphism $\varphi : F \to G_k$, inducing a map $\varphi' : E' \to \exp \mathfrak{E}'$ of exact sequences of groups which finally gives σ as composition

$$H_2(\Gamma;\mathbb{Z}) \simeq (F'\cap R)/(F,R) \to \exp((\mathfrak{f}'\cap\mathfrak{r})/[\mathfrak{f},\mathfrak{r}] \simeq H_2(\mathfrak{u}_k;k)) .$$

We shall show that $\varphi(F)$ is a S-arithmetic subgroup of G_k . This implies the proposition, since if $\varphi(F)$ is S-arithmetic in G_k , $\varphi(F')$ is S-arithmetic in G_k' by 7.2.4 and hence $\varphi(F'\cap R) = \varphi(F')\cap\exp(\mathfrak{r}/[\mathfrak{f},\mathfrak{r}])$ is S-arithmetic in $\exp((\mathfrak{f}'\cap\mathfrak{r})/[\mathfrak{f},\mathfrak{r}]) \simeq H_2(\mathfrak{u}_k;k)$.

So it remains to show that $\varphi(F)$ is S-arithmetic in G_k . The image C of Ω under

$$\Omega \to \exp \mathfrak{g}_k \to \prod_{v\in P-S} (\exp \mathfrak{g})_{k_v} = \prod_{v\in P-S} \exp(\mathfrak{g}_{k_v})$$

has compact closure - since the image of $\Omega \to \mathfrak{u}_k \to \prod_{v\in P-S} \mathfrak{u}_{k_v}$ does - , hence so does the group generated by it as follows immediately form 2.6.3 and the fact that the image of C under the projection $G_A \to G_{k_p}$ is contained in G_{O_p} for every $p \in V$ except finitely many. It follows that $\varphi(F)$ is contained in a S-arithmetic subgroup Δ of G_k . On the other hand, $\log \Omega \subset \mathfrak{u}_k$ contains a S-arithmetic subgroup of $\mathfrak{u}_k$ - as is easily seen by embedding $\mathfrak{u}_k$ into a Lie algebra of upper triangular matrices with zeroes on the diagonal and taking the log of matrices - , hence so does its image in $\mathfrak{g}^{ab}{}_k$, since $\mathfrak{u}_k$ maps onto $\mathfrak{g}^{ab}{}_k$. So the image of $\varphi(F)$ in $G^{ab}{}_k$ contains a S-arithmetic subgroup. This implies by induction on the descending central series G_i of G that the image of $G_i \cap \varphi(F)$ in G_i/G_{i+1} contains a S-arithmetic subgroup. So the image of $G_i \cap \varphi(F) \to (G_i/G_{i+1})_{SU\infty}$ is cocompact, hence so is the image of $\varphi(F) \to G_{SU\infty}$ which proves that $\varphi(F)$ is of finite index in Δ (see 7.1.5) □

As a corollary we obtain the two results we were after.

Fix $p \in S \cap P$. Let D be a maximal k_p-split subtorus of G and $H_2(\mathfrak{u})^0$ the weight zero subspace for the representation of D on $H_2(\mathfrak{u})$.

7.4.4 PROPOSITION *Let Γ be a S-arithmetic subgroup of the (not necessarily connected) solvable k-group G. Let Ω be a S-arithmetic subgroup of U normalized by Γ. Follow the map $H_2(\Omega;\mathbb{Z}) \to H_2(\mathfrak{u})$ by projection onto $H_2(\mathfrak{u})^0$. If the image of $H_2(\Omega;\mathbb{Z})$ in $H_2(\mathfrak{u})^0$ is a finitely generated $\mathbb{Z}\Gamma$-module, then $H_2(\mathfrak{u})^0 = 0$, i.e. zero is not a weight of D on $H_2(\mathfrak{u})$ (condition 2) of theorem 6.2.3).*

PROOF. The image A of $H_2(\Omega;\mathbb{Z}) \to H_2(\mathfrak{u})$ is a S-arithmetic subgroup of $H_2(\mathfrak{u})$ by 7.4.2 or 7.4.3, hence the closure $\bar{A}$ of A in $H_2(\mathfrak{u})_{k_p} = H_2(\mathfrak{u}_{k_p};k_p)$ is compact (7.1.4), hence so is the closure $\bar{B}$ of the image B of $H_2(\Omega;\mathbb{Z}) \to H_2(\mathfrak{u})^0_{k_p}$. On the other hand, the action of G_{k_p} on $H_2(\mathfrak{u})^0_{k_p}$ has compact orbits, since $G_{k_p}/(D\cdot U)_{k_p}$ is compact (7.1.7). By hypothesis B is finitely generated as a $\mathbb{Z}\Gamma$-module, hence $\bar{B}$ is compactly generated as a group with G_{k_p}-action. So $\bar{B}$ is compactly generated as a topological group, hence so is the vector space $H_2(\mathfrak{u})^0_{k_p}$ over k_p which is impossible unless $H^2(\mathfrak{u})^0 = 0$ □

7.4.5 COROLLARY *Let Γ be a S-arithmetic subgroup of the connected solvable k-group G. Suppose G is k-split and Γ is finitely generated. If $H_2(\Gamma;\mathbb{Z})$ is a finitely generated group, then, for every $p \in S$ condition 2) of theorem 6.2.3 holds.*

PROOF. By hypothesis $T = G/U$ is k-split, so $D \simeq T$ for every $p \in S$, $H_2(\mathfrak{u})^0$ is independent of $p \in S \cap P$ and G acts trivially on $H_2(\mathfrak{u})^0$.

Put $\Omega = \Gamma \cap U$. The map $H_2(\Omega) \to H_2(u) \to H_2(u)^o$ factors through $H_o(\Gamma/\Omega; H_2(\Omega))$. So $H_2(u)^o = o$ by 7.4.4 provided that $H_o(\Gamma/\Omega; H_2(\Omega))$ is a finitely generated abelian group. But this follows from our hypotheses and the next two lemmas □

7.4.6 LEMMA *Let* $A \to B \twoheadrightarrow C$ *be an exact sequence of groups and let* M *be a* $\mathbb{Z}B$*-module. Suppose* $H_i(C; H_j(A;M)$ *is finitely generated if* $i + j \leq 3$ *and* $j \leq 1$. *Then* $H_2(B;M)$ *is finitely generated iff* $H_o(C; H_2(A;M))$ *is finitely generated.*

PROOF by the Lyndon-Hochschild-Serre spectral sequence modulo the Serre class of finitely generated groups. See the following picture. Everything in the shaded area is finitely generated □

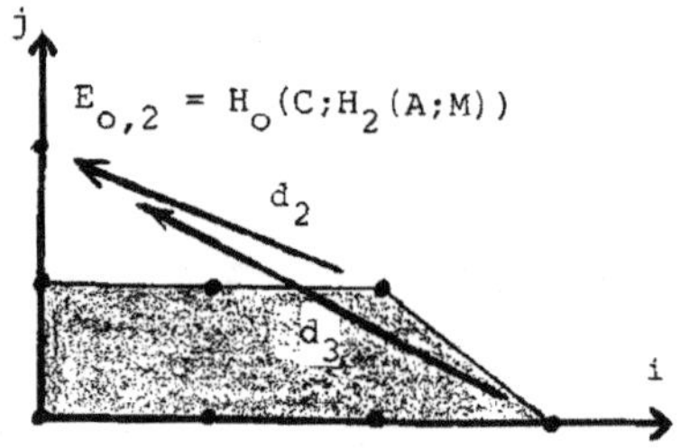

7.4.7 LEMMA *Let* Γ *be a S-arithmetic subgroup of a (not necessarily connected) solvable k-group. Put* $\Omega = \Gamma \cap U$. *Then* $H_i(\Gamma/\Omega; \mathbb{Z})$ and $H_i(\Gamma/\Omega; H_1(\Omega;\mathbb{Z})$ *are finitely generated for every* $i \in \mathbb{Z}$, *if* Γ *is finitely generated.*

PROOF. $Q = \Gamma/\Omega$ is isomorphic to the S-arithmetic (7.1.6) subgroup image$(\Gamma \to G/U)$ of G/U hence contains a finitely generated abelian subgroup of finite index (see 6.1.3). So $\mathbb{Z}Q$ is left noetherian.

This implies the first claim. If Γ is finitely generated, $(\Gamma \cap G^o)^{ab}$ is finitely generated abelian, hence finitely presented, so $(\Gamma \cap G^o)'$ is finitely generated as a normal subgroup of Γ (see 1.1.3 b)), hence so is the commensurable group $\Omega = \Gamma \cap U$ (7.2.10). It follows that $H_1(\Omega, \mathbb{Z}) = \Omega^{ab}$ is a finitely generated $\mathbb{Z}Q$-module, which implies the second claim, since $\mathbb{Z}Q$ is left noetherian □

7.4.8 Theorems 7.0.1 and 7.0.2 now follow from 6.2.3, 7.3.13, 7.2.9, 7.2.10, 7.4.4, 7.4.5 (sufficiency) and necessity from the following lemma and the basic result [14] of Bieri and Strebel, that a finitely presented solvable group is tame □

7.4.9 LEMMA *If* Γ *is a finitely presented group and* N *is a normal subgroup of* Γ *such that* Γ/N *is abelian, then* $H_i(N;\mathbb{Z})$ *is a finitely generated* $\mathbb{Z}\Gamma$*-module for* $i \leq 2$.

PROOF. If Γ is finitely presented, it is FP_2 (cf. [47] 2.1), i.e. the trivial $\mathbb{Z}\Gamma$-module $\mathbb{Z}$ has a resolution by projective $\mathbb{Z}\Gamma$-modules P_i

(*) $$\cdots\cdot \to P_2 \to P_1 \to P_0 \to \mathbb{Z}$$

which are finitely generated $\mathbb{Z}\Gamma$-modules for $i \leq 2$. Put $Q = \Gamma/N$, a finitely generated abelian group. The resolution (*) is also $\mathbb{Z}N$-projective, so $H_i(N;\mathbb{Z})$ is a subquotient of the finitely generated $\mathbb{Z}Q$-module $P_i \otimes_{\mathbb{Z}N} \mathbb{Z}$ for $i \leq 2$, hence is finitely generated itself, since $\mathbb{Z}Q$ is noetherian □

7.4.10 REMARK The same proof shows more generally the following result: *Let* Γ *be a group of type* FP_m , N *a normal subgroup of* Γ *such that* $\mathbb{Z}\Gamma/N$ *is left noetherian. Let* B *be a finitely generated* $\mathbb{Z}\Gamma$*-module, then* $H_i(N;B)$ *is a finitely generated* $\mathbb{Z}\Gamma$*-module for every* $i \leq m$ □

7.5 G NOT CONNECTED

In this section we finish the proof of theorem 7.5.2, see paragraph after 7.5.6.

7.5.1 Let G be a solvable k-group, Γ a S-arithmetic subgroup of G

and Δ a nilpotent normal subgroup of Γ such that Γ/Δ contains a finitely generated abelian subgroup Q of finite index, e.g. $\Delta = \Gamma \cap U$, $U = \mathrm{rad}_u G$, by 7.1.6 and 6.1.3.

7.5.2 THEOREM *Γ has a finite presentation iff the following two conditions hold:*

1) Δ^{ab} is a tame $\mathbb{Z}Q$-module.

2) $H_2(\Delta)$ is a finitely generated $\mathbb{Z}\Gamma$-module.

The main step in the proof is to see how Δ and $\Gamma \cap U$ are related, as follows.

7.5.3 PROPOSITION *Suppose $S \neq \emptyset$. Suppose Γ and Δ are as in 7.5.1. Then $\Delta \cap U$ is S-arithmetic in U and the center of $\Delta/\Delta\cap U'$ is of finite index and contains $\Delta\cap U/\Delta\cap U'$. In particular the image of Δ' in $\Delta/\Delta\cap U'$ is finite.*

PROOF. $\Gamma\cap U/\Delta\cap U$ is finitely generated, since it is a subgroup of Γ/Δ, hence is finite by the following lemma 7.5.4. For the other claims we may assume that $U' = 0$ and G is the algebraic hull of Δ. Note that U is contained in the algebraic hull of Δ, since $\Delta \cap U$ is S-arithmetic in U. Then G is nilpotent, hence $G^o = T \times U$ (see [18] 10.6(3)) is abelian. For the adjoint representation Ad of G on the Lie algebra $\mathfrak{g}$ of G^o the image Ad G is finite and consists of unipotent matrices, since G is nilpotent, hence Ad $G = \{e\}$, so G^o is central in G. The last claim follows from the following theorem of Schur: If the center of a group A is of finite index in A, then A' is finite (see [44] Theorem 4.12) □

7.5.4 LEMMA *Suppose $S \neq \emptyset$. Let Θ be a S-arithmetic subgroup of a unipotent k-group. A normal subgroup Λ of Θ such that Θ/Λ is finitely generated is of finite index.*

PROOF. First suppose $\Theta = o(S) \subset k$. Let $\Lambda \subset \Theta$ be such that Θ/Λ
is finitely generated. For every $p \in S$ there is an element $x_p \in o$
such that $\|x_p\|_p < 1$, by the finiteness of the class number. Then
$\bigcup_{n \in \mathbb{N}} x_p^{-n} \Lambda$ is a strictly ascending sequence of subgroups of Θ , which
contradicts the hypothesis that Θ/Λ is finitely generated abelian,
unless $x_p^{-1} \cdot \Lambda = \Lambda$. So Λ is a $\mathbb{Z}[x_p^{-1}]$-module for every $p \in S$, hence
a R-module with $R = \mathbb{Z}[x^{-1}]$, $x = \Pi x_p$, $p \in S$. So Θ/Λ is a R-module
and a finitely generated abelian group. It follows now easily by induc-
tion on the number of generators qua R-module that Θ/Λ is finite. In
fact, if R/dR is a cyclic R-module and a finitely generated group,
then $d \neq 0$ and it suffices to prove that R/dR is a torsion group,
i.e. that $R/dR \otimes \mathbb{Q} = \mathbb{Q}[x^{-1}]/d\mathbb{Q}[x^{-1}]$ is zero which is true since
$\mathbb{Q}[x^{-1}]$ is a field.

If now Θ is a S-arithmetic subgroup of a one dimensional unipotent
k-group U , then Γ is commensurable with a cyclic $o(S)$-module, hence
our result follows from the case $\Theta = o(S)$. Now apply induction on
dim U □

With notations as in 7.5.1 let Γ_1 be a subgroup of Γ of finite
index with the following properties: $G^o \supset \Gamma_1 \supset \Delta \cap G^o$ and $\Gamma_1/\Delta \cap G^o \subset Q$.

7.5.5 LEMMA *With notations as above, Γ_1 is tame iff Δ^{ab} is a tame
$\mathbb{Z}Q$-module.*

7.5.6 LEMMA *Suppose Γ is finitely generated. If $H_2(\Delta;\mathbb{Z})$ is a finitely generated $\mathbb{Z}\Gamma$-module, then, for every $p \in S$, condition 2) of
theorem 6.2.3 holds for Γ_1 .*

Note that with the proof of these two lemmas the proof of theorem
7.5.2 is finished: Γ is finitely presented iff Γ_1 is. Necessity
follows now from 7.5.5 and 7.4.9 and the fact that a finitely presented
solvable group is tame (see 7.3.10). Sufficiency follows from 6.2.3,
7.3.13, 7.5.5 and 7.5.6.

PROOF OF 7.5.5. The first three paragraphs of this proof will be used in the proof of 7.5.6, too.

Since $\Gamma_1 \subset G^o$ and $\Gamma_1/\Delta\cap G^o$ is abelian, we have $\Gamma_1' \subset \Delta \cap U$, so $\Gamma_1/\Delta\cap U$ is abelian.

$\Gamma/\Gamma\cap U$ is isomorphic to the S-arithmetic (7.1.6) subgroup image($\Gamma \to G/U$) of the group G/U whose connected component is a torus. So $\Gamma/\Gamma\cap U$ contains a finitely generated abelian subgroup of finite index (see 6.1.3). In particular, every subgroup of $\Gamma/\Gamma\cap U$ is finitely generated. Hence $\mathbb{Z}\Gamma_1/\Delta\cap U$ is a noetherian ring and hence $\mathbb{Z}\Gamma/\Delta\cap U$ is left noetherian.

All the groups in the following sequence

(7.5.7) $\qquad (\Delta\cap U)' \subset \Delta'\cap U' \subset \Delta'\cap U \subset \Delta'$, $\Delta\cap U'$

are S-arithmetic subgroups of U' by 7.5.3 and 7.2.4.

For the proof of 7.5.5 itself we may assume that Γ is finitely generated, since either Γ_1 is tame, hence finitely generated (7.3.10, 7.2.9) or Δ^{ab} is a finitely generated $\mathbb{Z}Q$-module, hence the inverse image of Q under $\Gamma \to \Gamma/\Delta$ - a subgroup of Γ of finite index - is finitely generated (see the second part of the proof of 7.2.9).

We shall frequently use the following obvious equivalence. Let $A \to A'$ be a map of $\mathbb{Z}B$-modules, B a finitely generated abelian group, whose kernel and cokernel are finitely generated abelian groups. Then A is a tame $\mathbb{Z}B$-module iff A' is a tame $\mathbb{Z}B$-module.

Now put $\Lambda = \Delta\cap U/\Delta\cap U'$. Then Λ is a $\mathbb{Z}Q$-module since Δ centralizes $\Delta \cap U$ mod $\Delta \cap U'$ by 7.5.3. Now Δ^{ab} is a tame $\mathbb{Z}Q$-module iff Λ is a tame $\mathbb{Z}Q$-module, because in the following diagram

$$\Delta\cap U/\Delta\cap U' \leftarrow \Delta\cap U/\Delta'\cap U' \to \Delta/\Delta'$$

of $\mathbb{Z}Q$-modules the left hand arrow is surjective and has finite kernel by 7.5.7 and the right hand arrow has finite kernel by 7.5.7 and the cokernel is finitely generated qua group, since it is the image of the subgroup $\Delta/\Delta\cap U$ of $\Gamma/\Gamma\cap U$.

Λ is a $\mathbb{Z}\Gamma_1^{ab}$-module, since $\Gamma_1' \subset \Delta \cap U$ centralizes $\Delta \cap U$ mod $\Delta \cap U'$. The $\mathbb{Z}Q$-module Λ is tame iff the $\mathbb{Z}\Gamma_1^{ab}$-module Λ is tame since the natural map $\Gamma_1^{ab} \to Q$ has finite cokernel and its kernel $\Delta \cdot \Gamma_1'/\Gamma_1'$ acts trivially on Λ. Finally Λ is a tame $\mathbb{Z}\Gamma_1^{ab}$-module iff $\Gamma_1'^{ab}$ is a tame $\mathbb{Z}\Gamma_1^{ab}$-module since the arrows in the diagram

$$\Gamma_1'/\Gamma_1'' \to \Gamma\cap U/\Gamma\cap U' \leftarrow \Delta\cap U/\Delta\cap U'$$

of $\mathbb{Z}\Gamma_1^{ab}$-modules have finite kernel and cokernel by 7.2.10, 7.2.4 and 7.5.3 □

PROOF OF 7.5.6. Put $\Omega = \Delta \cap U$. Look at the spectral sequence of

$$\Omega \rightarrowtail \Delta \to \Delta/\Omega$$

and apply the argument of 7.4.6 replacing the Serre class of finitely generated groups by the Serre class of finitely generated $\mathbb{Z}\Gamma_1$-modules which is possible since $\mathbb{Z}\Gamma/\Omega$ is noetherian (second paragraph of the proof of 7.5.5). So $H_0(\Delta/\Omega; H_2(\Omega))$ is a finitely generated $\mathbb{Z}\Gamma$-module iff $H_2(\Delta)$ is, provided $H_i(\Delta/\Omega; H_j(\Omega))$ is a finitely generated $\mathbb{Z}\Gamma$-module for $i + j \leq 3$ and $j \leq 1$. Now $H_0(\Omega) = \mathbb{Z}$ and $H_1(\Omega)$ is a finitely generated $\mathbb{Z}\Gamma$-module by 7.2.10, 7.5.3 and 7.5.7, since Γ is finitely generated by hypothesis. So $H_i(\Delta/\Omega; H_j(\Omega))$ is a finitely generated $\mathbb{Z}\Gamma$-module for $j \leq 1$ and arbitrary i by 7.4.10 since $\mathbb{Z}\Gamma/\Omega$ and $\mathbb{Z}\Gamma/\Delta$ are left noetherian rings, hence Γ/Ω is FP_m for every m. So $H_0(\Delta/\Omega; H_2(\Omega))$ is a finitely generated $\mathbb{Z}\Gamma$-module, by hypothesis.

For $p \in S \cap P$, let $H_2(\mathfrak{u})^0$ be the subspace of D-fixed vectors, D the maximal k_p-split subtorus of G/U. Now G^0 acts on $H_2(\mathfrak{u})^0$ by semisimple endomorphisms and, since Δ is nilpotent, Δ acts unipotently on $\mathfrak{u}$, since $\Delta \cap U$ is a S-arithmetic hence Zariski-dense in U, hence Δ acts on $H_2(\mathfrak{u})^0$ by unipotent endomorphisms, so $G^0 \cap \Delta$ acts trivially on $H_2(\mathfrak{u})^0$. Similarly, for the representation of $\Delta/G^0\cap\Delta$ on $H_2(\mathfrak{u})^0$ the image is finite and consists of unipotent endomorphisms, hence is trivial. So the map

$$H_2(\Omega) \to H_2(\mathfrak{u}) \to H_2(\mathfrak{u})^0$$

factors through $H_o(\Delta/\Omega;H_2(\Omega))$, hence its image is a finitely generated $\mathbb{Z}l$-module, which implies 7.5.6 by 7.4.4 □

APPENDIX, LINEAR INEQUALITIES

Here we recall two results about the existence of integer valued linear forms having positive values on certain subsets.

Let V be a finite dimensional vector space over $\mathbb{Q}$. Let T be a finite subset of V. Define $V_{\mathbb{R}} = \mathbb{R} \otimes_{\mathbb{Q}} V$. Then we have

A.1 PROPOSITION *1) There is a $\mathbb{Q}$-linear form ℓ on V such that $\ell|T > 0$ iff there is a $\mathbb{R}$-linear form ℓ on $V_{\mathbb{R}}$ such that $\ell|T > 0$.*
2) There is a $\mathbb{Q}$-linear form $\ell \neq 0$ on V such that $\ell|T \geq 0$ iff there is a $\mathbb{R}$-linear form $\ell \neq 0$ on $V_{\mathbb{R}}$ such that $\ell|T \geq 0$.

PROOF. 1) The set of $\ell \in (V_{\mathbb{R}})^* = (V^*)_{\mathbb{R}}$ satisfying $\ell|T > 0$ is open, hence contains an element of the dense subset V^* if non-empty.
2) Necessity is trivial. Suppose there is an $\ell_o \neq 0$ in $(V_{\mathbb{R}})^*$ such that $\ell_o|T \geq 0$. Set $T_1 = \{t \in T ; \ell_o(t) = 0\}$, $T_2 = \{t \in T ; \ell_o(t) > 0\}$. For the vector subspaces $(T_{1,\mathbb{R}})^{\perp} = \{\ell \in V_{\mathbb{R}}^* ; \ell(t) = 0 \text{ for } t \in T_1\}$ and $T_1^{\perp} = \{\ell \in V^* ; \ell(t) = 0 \text{ for } t \in T_1\}$ we have $(T_1^{\perp})_{\mathbb{R}} = (T_{1,\mathbb{R}})^{\perp}$ because $(T_1^{\perp})_{\mathbb{R}} \subset (T_{1,\mathbb{R}})^{\perp}$ and both have dimension $\dim V - \dim \operatorname{span} T_1$. Now $\{\ell \in (T_{1,\mathbb{R}})^{\perp} ; \ell|T_2 > 0\}$ is an open non-empty subset of $(T_{1,\mathbb{R}})^{\perp}$, therefore contains an element of the dense subset $T_1^{\perp}$ □

A.2 COROLLARY *The following three statements are equivalent.*
1) There is no linear form $\ell \in V^*$ such that $\ell|T > 0$.
2) There is a non trivial linear relation $\sum_{t \in T} n_t \cdot t = 0$ with integer coefficients $n_t \geq 0$.
3) 0 is in the convex hull of T in $V_{\mathbb{R}}$.

PROOF. 1) $\Leftrightarrow$ 3) By the theorem on separating hyperplanes, 3) is equivalent to the following statement. There is no linear form $\ell \in V_{\mathbb{R}}^{*}$ such that $\ell|T > 0$. This is equivalent to 1) by A.1.
2) $\Leftrightarrow$ 3) Define $X = \{(x_t)_{t\in T} \in \mathbb{Q}^T ; \Sigma x_t \cdot t = 0\}$, $Y = X^*$ and $S \subset Y$ the set of linear forms p_t, $t \in T$, on X, where p_t is the projection onto the t-component. Now 0 is in the convex hull of T in $V_{\mathbb{R}}$ iff there is a non-zero $x \in X_{\mathbb{R}} = Y_{\mathbb{R}}^{*}$ such that $\langle x,s\rangle = \langle s,x\rangle \geq 0$ for every $s \in S$ iff there is a non-zero $x \in X$ such that $\langle x,s\rangle \geq 0$ for every $s \in S$, by A.1, iff there are integers $n_t \geq 0$, not all of them zero, such that $\Sigma\, n_t \cdot t = 0$ □

A subset T of a real vector space is called *positively dependent* if zero is in the convex hull of T, equivalently if there are non-negative real numbers a_t, $t \in T$, all but a finite number of them zero, but not all of them zero, such that $\sum_{t\in T} a_t \cdot t = 0$.

A.3 REMARK If T is a finite subset of a (finite dimensional) vector space V over $\mathbb{Q}$, then T is positively dependent in $V_{\mathbb{R}}$ iff there are non-negative rational (integer) numbers a_t, $t \in T$, not all of them zero, such that $\sum_{t\in T} a_t \cdot t = 0$.

REFERENCES

1 ABELS, H.: Kompakt definierbare topologische Gruppen. Math. Ann. 197 (1972) 221 - 233

2 - Normen auf freien topologischen Gruppen. Math. Zeitschr. 129 (1972) 25 - 42

3 - An example of a finitely presented solvable group. In: Homological Group Theory, Proceedings Durham 1977, ed. C.T.C. Wall, London MS Lecture Note 36 (1979) 205 - 211

4 - Finite presentability of S-arithmetic groups. Proc. Conf. Groups, St. Andrews, 1985

5 ABELS, H. and P. ABRAMENKO: On the homotopy type of subcomplexes of Tits buildings. Preprint

6 ABELS, H. and K.S. BROWN: Finiteness properties of solvable S-arithmetic groups: An example. J. Pure Applied Algebra 1987

7 ÅBERG, H.: Bieri-Strebel valuations (of finite rank). Proc. London MS (3) 52 (1986) 269 - 304

8 - Solvable groups and FP_n . Preprint

9 BEHR, H.: Über die endliche Definierbarkeit verallgemeinerter Einheitengruppen. II. Inv. math. 4 (1967) 265 - 274

10 - Finite presentability of arithmetic groups over global function fields. Groups St. Andrews 1985. Proc. Edinburgh Math. Soc. 30 (1987) 23 - 39

11 BIERI , R.: Homological dimension of discrete groups. Queen Mary College. Mathematics Notes 1976

12 - A connection between the integral homology and the centre of a rational linear group. Math. Z. 170 (1980) 263 - 266

13 BIERI, R. and R. STREBEL: Almost finitely presented soluble groups. Comm. Math. Helv. 53 (1978) 258 - 278

14 BIERI, R. and R. STREBEL: Valuations and finitely presented metabelian groups. Proc. London MS (3) 41 (1980) 439 - 464

15 BOREL, A.: Arithmetic properties of linear algebraic groups. Proc. Int. Congr. Mathematicians Stockholm 1962, Uppsala 1963, 10 - 22 = Coll. Papers II 331 - 343

16 - Some finiteness properties of adele groups over number fields. Publ. Math. Inst. Hautes Etud. Sci. 16 (1963) 5 - 30 = Coll. Papers II, 305 - 330

17 - Density and maximality of arithmetic subgroups. J. Reine Angew. Math. 224 (1966) 78 - 89 = Coll. Papers II, 635 - 646

18 - Linear algebraic groups. Notes by H. Bass. Benjamin 1969

19 BOREL, A. and HARISH-CHANDRA: Arithmetic subgroups of algebraic groups. Ann. Math. (2) 75 (1962) 485 - 535 = Borel, Coll. Papers II 236 - 286

20 BOREL, A. and J.P. SERRE: Corners and arithmetic groups. Comm. Math. Helv. 48 (1973) 436 - 491 = Borel, Coll. Papers III, 244 - 299

21 - Cohomologie d'immeubles et de groupes S-arithmetiques. Topology 15 (1976) 211 - 232 = Borel, Coll. Papers III, 439 - 460

22 BOREL, A. and J. TITS: Groupes rêductifs. Publ. Math. IHES 27 (1965) 55 - 152 = Borel, Coll. Papers II, 424 - 520

23 BOURBAKI, N.: Topologie Gênêrale. Chap. I - IV (1971)

24 - Groupes et algêbres de Lie. Chap. II, III (1972) Chap. IV - VI (1968), Chap. VII, VIII (1975)

25 BROWN, K.S.: Finiteness properties of groups. J. Pure Applied Algebra 1987

26 DEMAZURE, M. and P. GABRIEL: Groupes algébriques, I. 1970 Masson, Paris; North-Holland, Amsterdam

27 DYER, J.L.: A nilpotent Lie algebra with nilpotent automorphism group. Bull. AMS 76 (1970) 52 - 56

28 GRAEV, M.J.: Free topological groups. Izvestija Akad. Nauk SSSR, Ser. Mat. 12 (1948) 279 - 324, Engl. translation in AMS Transl. Ser 1,8 (1962) 305 - 364

29 - Theory of topological groups I. Norms and metrics on groups. Complete groups. Free topological groups (russ.) Uspehi Mat. Nauk 5,2 (36), (1950) 1 - 56

30 HALL, P.: A contribution to the theory of groups of prime power order.

Proc. London MS (2) 36 (1933) 29 - 95

31 HALL, P.: Finiteness conditions for soluble groups. Proc. London Math. Soc. (3) 4 (1954) 419 - 436

32 HILTON, P. and U. STAMMBACH: A course in homological algebra. Springer GTM 4 (1971)

33 HOLZ, S.: Endliche Identifizierbarkeit von Gruppen. Thesis Bielefeld 1985

34 KAPLANSKY, I.: An introduction to differential algebra, Act. Sci. Ind. 1251 (1957) Hermann, Paris

35 KARRASS, A. and D. SOLITAR: Subgroups of HNN-groups and groups with one defining relation. Can. J. Math. 23 (1971) 627 - 643

36 KNESER, M.: Erzeugende und Relationen verallgemeinerter Einheitengruppen. Crelle's Journal 214/215 (1964) 345 - 349

37 LAMBE, L.A. and S.B. PRIDDY: Cohomology of nilmanifolds and torsion-free nilpotent groups. Trans. AMS 273 (1982) 39 - 55

38 LAZARD, M.: Sur les groupes nilpotents et les anneaux de Lie. Ann. Sci. ENS (3) 72 (1954) 101 - 190

39 MALCEV, A.: On a class of homogeneous spaces. Izv. Akad. Nauk SSSR Ser. Mat. 13 (1949) 9 - 32; Engl. transl. Math. USSR-Izv. 39 (1951), also in AMS Translations Ser. 1, vol. 9, Lie groups (1962) 277 - 307

40 MARKOV, A.A.: On free topological groups. Izv. Akad. Nauk SSSR, Ser. Mat. 9 (1945) 3 - 64, Engl. translation in AMS Transl. 1,8 (1962) 195 - 272

41 PICKEL, P.F.: Rational cohomology of nilpotent groups and Lie algebras. Comm. Alg. 6 (4) 409 - 419 (1978)

42 QUILLEN, D.: Higher algebraic K-theory I. Proceedings Seattle 1973. Springer Lecture Note 341, 85 - 147

43 RAGHUNATHAN, M.S.: A note on quotients of real algebraic groups by arithmetic subgroups. Invent. math. 4 (1968) 318 - 335

44 ROBINSON, D.J.S.: Finiteness conditions and generalized soluble groups I, Springer Ergebnisse Bd. 62 (1972)

45 SAMUEL, P.: On universal mappings and free topological groups. Bull. AMS 54 (1948) 591 - 598

46 SERRE, J.P.: Lie algebras and Lie groups. Harvard Lecture Notes. Benjamin 1965

47 SERRE, J.P.: Cohomologie des groupes discrets. Prospects in Mathematics. Annals of Math. Studies 70, Princeton University Press 1971, 77 - 169

48 - Arbres, amalgames, SL_2 . Astérisque 46 (1977)

49 STREBEL, R.: Finitely presented soluble groups. In: Group Theory. Essays for Philip Hall. Ed. by K.W. Gruenberg and J.E. Roseblade. Acad. Press 1984

50 STUHLER, U.: Homological properties of certain arithmetic groups in the function field case. Inv. math. 57 (1980) 263 - 281

51 SULLIVAN, D.: Infinitesimal computations in topology. Publ. Math. IHES 47 (1977) 269 - 331

52 TITS, J.: Liesche Gruppen und Algebren. Vorlesung Bonn 1965

53 WALL, C.T.C.: Finiteness conditions for CW-complexes. Ann. of Math. 81 (1965) 56 - 69

54 - Finiteness conditions for CW-complexes II. Proc. Royal Soc. A 296 (1966) 129 - 139

55 WARFIELD, R.B.: Nilpotent groups. Springer Lecture Note 513 (1976)

56 WEIL, A.: Basic number theory. 3rd edition. Springer 1974

57 WITT, E.: Treue Darstellungen Liescher Ringe. J. reine angew. Math. 177 (1937) 152 - 160

List of Symbols

$\mathbb{A}$ [ring of adeles] 7.1.1

$[\mathfrak{a},\mathfrak{b}]$ 2.1

$\ldots^{ab}$ [abelianization] 2.1, 2.3.5

(A,B) 2.3.1

$\mathbb{A}_k$ [ring of adeles] 7.1.1

B_+ , B_- [Borel subgroup] 6.4.1

C [subsemigroup of $\mathrm{Hom}(Q,\mathbb{R})$] 4.1

$\mathcal{C}$ [ordered set of C's with $0 \notin C$] 4.2.8

conv [convex hull] Appendix

C_t 4.1.2

Δ [set of simple roots] 6.4.1

$\Phi = \Phi(T,G)$ [set of roots] 6.3.1

Φ_+ [set of positive roots] 6.4.1

$F(X)$ [free group with basis X] 1.1

$\mathfrak{g}'$ [commutator Lie algebra] 2.1

$G' = (G,G)$ [commutator subgroup] 2.3.5

$G_\infty = G_{V-P}$ 7.1.2

G^o [connected component of G] 7.5

$G_B , G_{\mathcal{O}(S)}$ 7.1.2

$G_S = \prod_{v \in S} G_v$ 7.1.2

$G_{S\cup\infty} = G_{S\cup(V-P)}$ 7.1.2

$G_v = G_{k_v}$ 7.1.2

$H = \coprod_{\mathcal{C}} Q \ltimes N^C$ 4.2.8

$H_i(\mathfrak{g}) = H_i(\mathfrak{g};k)$ 5.2.1

$H_i(G) = H_i(G;\mathbb{Z})$ 5.1.3

$j_S : G_k \to G_S$ 7.1.2

$k_{\mathfrak{p}}, k_v$ [completion of k with respect to $\mathfrak{p}$, v] 7.1.1

L [set of lines in $\mathrm{Hom}(Q,\mathbb{R})$] before 4.4.12

$\mathfrak{m}$ [maximal ideal of $\mathcal{O}$] 2.6

$M = \coprod_{\mathcal{C}} N^C$ 4.2.8

mod_K [module of a local field K] 1.2

$\nu = -\log \circ \mathrm{mod}_K$ 3.1.4

ν_* 3.1.4

N_+ , N_- [unipotent radical of B_+, B_-] 6.4.1

N^C 4.1

$\mathcal{O}$ [ring of integers] 2.6, 6.1.1, 7.1.1

$\mathcal{O}_{\mathfrak{p}}$ [ring of $\mathfrak{p}$-adic integers] 6.1.1, 7.1.1

$\mathcal{O}_v$ 7.1.1

$\mathcal{O}(S)$ [S-arithmetic ring] 6.1.1, 7.1.1

INDEX